NO BREATHING ROOM

The Aftermath of Chernobyl

Grigori Medvedev

Translated from the Russian by Evelyn Rossiter

WITH AN INTRODUCTION BY DAVID R. MARPLES

BasicBooks
A Division of HarperCollinsPublishers

Manque d'Oxygene is published with the permission of Éditions Albin Michel.

French translation: Copyright © 1993 Éditions Albin Michel S. A., Paris

English translation: Copyright © 1993 by BasicBooks, A Division of HarperCollins Publishers, Inc.

Designed by Barbara DuPree Knowles and Ellen Levine

LIBRARY OF CONGRESS CATALOGING-IN-PUBLICATION DATA
Medvedev, Grigori.
  [Bez kisloroda. English]
  No breathing room : the aftermath of Chernobyl / Grigori Medvedev ; translated from the Russian by Evelyn Rossiter ; with an introduction by David R. Marples.
    p. cm.
  Includes index.
  ISBN 0–465–05114–6 (cloth)
  ISBN 0–465–05115–4 (paper)
  1. Nuclear industry—Soviet Union. 2. Public relations—Nuclear industry—Soviet Union. 3. Official secrets—Soviet Union. 4. Chernobyl Nuclear Accident, Chernobyl, Ukraine, 1986—Social aspects. I. Title.
HD9698.S652M4313 1993
363.17'99'0947714—dc20   91–59061
                                              CIP

94 95 96 97 CC/RRD 9 8 7 6 5 4 3 2 1

Contents

Introduction by David R. Marples
1

PART ONE
Long Before Chernobyl
31

PART TWO
After Chernobyl
105

PART THREE
Frontal Assault
153

Postscript
197

Index
199

Introduction

AS A NUCLEAR ENGINEER and a departmental chief in the Directorate for Nuclear Energy in the former Soviet Union, Grigori Medvedev has had extensive experience in the nuclear power industry and was one of the experts summoned to deal with the aftereffects of the Chernobyl disaster in the summer of 1986. In his 1991 publication *The Truth About Chernobyl** (originally called *Chernobyl Notebook*) Medvedev provided a grimly realistic picture of the scene at the nuclear plant in the immediate aftermath of the tragedy. It was the first full-length eyewitness account of these events to appear in English, and it was notably free of the descriptions of heroic acts that characterized previous articles by Soviet writers on the subject of Chernobyl. It portrayed a group of overconfident and yet alarmingly ignorant senior officials who failed to comprehend and then refused to acknowledge the enormity of what had happened. Until Medvedev's book appeared, it had not been possible to gauge accurately just how the officials on the spot could have so seriously misread the situation.

The current book, *No Breathing Room*, might be seen as in the same vein: it also provides revelations about the operation of the Soviet nuclear industry and its inherent secrecy. Yet it also goes further than its predecessor in that it provides us with an in-depth analysis of the Soviet bureaucracy that developed in the Brezhnev period. This

* Basic Books, 1991.

book is a personal account of one man's struggle against the system, and against the censor in particular; it is the story of an honest and straightforward man whose work was constantly impeded and withheld from publication by government officials, even during the period known as "*glasnost* and *perestroika*." Medvedev describes the Soviet system as an "octopus," with each tentacle representing a different ministry. Although the Soviet Union has collapsed, Medvedev's book remains both pertinent and vital, for many aspects of the government structure have remained. Moreover, there have been at least three major accidents in the post-Soviet nuclear power industry—or related to that industry—in 1991–92: a mishap that necessitated the shutdown of reactor No. 2 at Chernobyl in October 1991; a radiation leakage at the Leningrad nuclear power plant in March 1992; and a dangerous fire that spread Chernobyl-deposited radiation in the southern regions of the Gomel province of Belarus in May 1992. Each of these three incidents could have resulted in a serious catastrophe; each testifies to the timeliness of Medvedev's new book; and none received the sort of publicity that it warranted in the press of the new state in which it occurred.

Many of the departments and ministries portrayed in this book may be unfamiliar to the general reader. They were barely known even to the Soviet public. It is useful, therefore, in introducing this book to portray the society in which Medvedev lived and worked, to convey the background for his struggle against the censor and against the Soviet authorities. For although Medvedev held a sensitive position in the nuclear industry, it is clear that he felt a compulsion to write: about the cover-ups of previous accidents; about the major explosion in the

Chelyabinsk region in the late 1950s; and, above all, about Chernobyl, the disaster whose consequences continue to dog post-Soviet society today. Medvedev sees a natural progression in these events. In his view, Soviet society was hurtling toward a nuclear abyss. His mission, as a member of the nuclear industry, was to try to prevent disaster by exposing the flaws and dangers of the system through the printed word.

THE "PERIOD OF STAGNATION"

At the heart of the Soviet system lay the Communist Party of the Soviet Union (CPSU), which according to the Brezhnev Constitution of 1977 had been officially assigned the leading role in Soviet society. As most people realized, the party had in fact wielded most of the power since the revolution; Brezhnev only confirmed it in black and white. Alongside the party leadership was the government—Lenin had referred to this as "Soviet power"—whose authority was never as strong as it sounded. The Soviet economy was subjected to partial and incomplete reforms at a bewildering rate between 1929 and 1966. Institutions and ministries were revamped, abolished, amalgamated. By 1964 life had become so insecure that the KGB decided to remove Khrushchev with a deft coup, as efficient and clinical as the August 1991 coup was clumsy and farcical. Like Gorbachev, Khrushchev was taking a vacation in the Crimea. His movements were carefully screened by the KGB leaders who engineered his downfall—which was approved by a carefully packed Central Committee meeting—and a faceless bureaucrat, Leonid Ilich Brezhnev, was installed in his place.

The Brezhnev era in the history of the Soviet Union was remarkable for its longevity. Mikhail Gorbachev described it as the "period of stagnation," and although this was not strictly accurate—Brezhnev, like his predecessors, began his tenure as General Secretary with a period as a reformer—it seemed a fitting image for an aging and geriatric leadership. Nothing symbolized the internal decay of the Soviet Union after 1975 more than the stumbling leader in his ill-fitting clothes, using cue cards to make his speeches and receiving medals for feats that he could not possibly have accomplished. The same Brezhnev who was seen at international conferences reading a speech twice because he had shuffled his cue cards from front to back was awarded prizes for literature. In 1987 a young Intourist guide from Moscow informed me that the most tedious task of his college days had been perusing eighteen volumes of Brezhnev's works. Yet the list of Brezhnev's achievements, from war hero to scientific innovator, went on and on.

This was an era of superficial greatness for the Soviet Union. Its economy was aptly described by a Western scholar as that of a first-world military power with a third-world economy. A highly centralized state, the Soviet Union could dispatch unmanned capsules into space yet, since the Khrushchev period, had been unable to harness its vast resources to feed its own people or to provide them with basic goods. Nevertheless, the system seemed greater than its parts. Leaders might come and go, but the economy spluttered on, under what was described subsequently as an "administrative-command system." Official statistics reported that the Soviet Union was the world's leading producer of oil and that coal production had achieved new breakthroughs in Siberia and the Far East. Above all, there were external appearances to be kept up, and none more so than the annual show of military hardware each November 7, the anniversary of the 1917 revolution. One wondered why the world needed to witness this alarming display of rockets and tanks, but traditions die hard.

To be a party apparatchik in this era was to have a career for life. One could rarely be dismissed and the privileges were considerable. One had access to special stores, special goods, country houses, trips abroad, and, at the highest levels, personal chauffeurs who would transport one in black Volgas with darkened windows, so that as a Soviet leader one need never face the stares of the workers that the party purported to represent. The elite was given a name—the *nomenklatura*—and, like other sectors of society, it became institutionalized, bureaucratized. The satiated superpower had, according to Khrushchev, attained socialism by 1961. The quest then focused on the utopia of communism. Khrushchev, an affable peasant who seemed to introduce reforms on momentary whims, considered 1980 to be the likely date for achieving utopia. That date was fast approaching, however, and a disillusioned public began to recognize that Soviet society was unlikely to provide them with the necessities of life: an apartment, a car, or a refrigerator that worked properly. In fact, daily life depended upon access to the black market, upon standing in line for hours at a time without even knowing what goods were being dispensed at the head of the line.

Although Brezhnev may have turned out to be more tenacious and cunning than the KGB anticipated, he eventually obliged by providing the country with stability and job security. Even the root of his name, *bereznyi*, is

Russian for *careful.* There were occasional moments of reaction—against recalcitrant Georgian Communists, against Czechoslovak reformers at the time of the Prague Spring in 1968—and, eventually, a disastrous invasion of Afghanistan in late 1979. But these events spanned almost the entire Brezhnev period. In the years between these events, the country was run by what Yurii Shcherbak, the Ukrainian Minister of the Environment, has described as "faceless armchair bureaucrats" who operated ministries from comfortable offices in Moscow and apparently felt not the slightest inclination to visit the regions that their decisions affected. These were men—and most all of them *were* men—who often cared little for their work. Indeed, some of their decisions appear in retrospect so ludicrous that one cannot conceive of any real attention to duty. The economy was run once again by ministries, each competing for its own power base, jealously guarding its territory against intruders. The coal ministry would be at loggerheads with the gas ministry; or ferrous metallurgy would struggle for increased investment at the expense of machine tools. Ultimately each ministry needed the others to survive because of the chaotic pattern of supply within the country.

There was, of course, some opposition to this state of affairs. It consisted of what Georgii Arbatov once referred to as a few hundred troublemakers who refused to participate in the Soviet system. Dissidents. In almost all cases, it was alleged, such "troublemakers" were operating at the behest of foreign powers. As long as they did not break the law, they were left alone. However, most of them were being encouraged by Western news agencies and soon took actions that contravened the criminal code. How else could the KGB react but by taking action to protect innocent citizens from these recalcitrants? Many of the dissidents were declared to be insane. They opposed the party, which represented the Soviet people, and were thus clearly unbalanced. Former war heroes, among them General Petro Grigorenko, were dispatched to insane asylums. The fact that these dissidents embraced a wide spectrum of Soviet society was studiously ignored. There were Marxist-Leninists (Roy Medvedev, Petro Grigorenko), Ukrainians (Vyacheslav Chornovil), Jews (Anatolii Shcharansky), Russians (Aleksandr Solzhenitsyn), writers (Aleksandr Tvardovsky, a portrait of whom is provided by Medvedev), and prominent scientists (Andrei Sakharov). People whose names would become household words in Gorbachev's Soviet Union were regarded as parasites under the Brezhnev regime.

THE SOVIET NUCLEAR POWER PROGRAM

In this rigid, stultifyingly bureaucratized society was launched one of the world's most ambitious nuclear power programs, commencing in the 1970s. It was the world that Grigori Medvedev inhabited and struggled against. A world of technocrats, secrecy, and censorship. It began with the military authorities. Chernobyl was a second-generation graphite-moderated (RBMK) reactor. Such reactors were originally part of the military economy—that is, geared to the production of plutonium and tritium for the nuclear weapons program. In the early 1970s former military reactors were adapted to the civilian program, with the pioneer station being the nuclear power station near Leningrad. The RBMK was and remains an exclusively Soviet reactor. It was never exported to Eastern Europe or to such friendly nations as Cuba and Vietnam. It was constructed without a concrete covering so that it could be refueled on line in its system of some 1,600 channels. Its efficiency remained limited to its ability to stay on line, however.

Because of the military nature of this reactor, it was not administered by the Ministry of Power, the body that ran the Soviet nuclear energy program prior to the Chernobyl disaster. Instead, the Ministry of Medium Machine Building (Minsredmash), which is described at length in *No Breathing Room,* had overall control. This ministry, whose delightfully nebulous name concealed its true nature, was run by aged bureaucrats and led by an octogenarian minister who was gracefully retired after Chernobyl. It has been argued that the Soviet military program was run more efficiently than the civilian economy, and certainly the military received priority in terms of investment. Yet once the nuclear plants became part of the civilian economy, they were treated like any other Soviet industry. They were subject to the same plans, the same supply problems, and the same shoddy construction; corners were cut whenever possible. A researcher at the US Department of State once told me of a revealingly worded advertisement he had seen in a Central Asian newspaper prior to the Chernobyl disaster: "Wanted: Operators for Nuclear Power Stations in the Ukraine. No Experience Necessary."

Why was there such an expansive program of construction in the 1970s and early 1980s? The Soviet Union had enjoyed a relative boom period in its oil and gas industry in the 1970s, but production peaked by the middle of that decade. In addition, world oil prices began to fall. In the coal industry, underground mining became more complex than in the peak production periods of the early 1970s. Mines were deeper, seams narrower, and the coal was difficult to extract using traditional technology. The Soviet coal industry experienced the highest accident rate in the world. According to one estimate taken from the Donbass coal field in the Ukraine—the largest coal field in the former Soviet Union—four miners were killed on average for every million tons of coal mined. Production had fallen, and the Soviet authorities redirected investment in the coal industry from the heavily exploited underground coal fields of Donbass to the more profitable mines of the Far East, in which coal is contained close to the surface. The problem—and this is not unique to the former Soviet Union—was that energy resources were located in remote areas, far from the main population and industrial centers. The Brezhnev regime sought an alternative: an industry that could provide a guaranteed supply of electricity in the heavily industrialized and most populated regions of the country. Nuclear energy seemed to provide a ready solution.

For years Soviet information about domestic industry had been subject to distortion. In 1948, when Stalin and Molotov rejected the Marshall Plan of US economic aid to war-torn Europe, many observers believed that the main reason was not distrust of US intentions, but unwillingness to publish accurate economic data about the state of the country, which the plan required. There were more than one hundred "closed cities," many of which were associated with some form of military installation or industry. But within Soviet civilian industries there were also very few disclosures of serious problems. Accidents were concealed whenever possible. It is hardly surprising, therefore, that the Soviet nuclear power industry was subject to comparable conditions. It had begun as a military industry, after all, and the military complex still retained control over some of the graphite stations, since they could readily be converted for military purposes. The 1970s then saw a rapid expansion of Soviet industry that was reminiscent of the emphasis on quantity over quality. It was reminiscent of the time of Stakhanov, in the mid-1930s, when an obscure Ukrainian miner overfulfilled his work quota under artificial conditions and was held up as a model for Soviet industry to emulate. At the same time, the outside world received little overt information about the inherent quality and supply predicaments of the factories, or about the high accident rate within them.

The history of the Soviet nuclear industry is not a happy one. In the late 1950s an explosion at a nuclear waste dump near Chelyabinsk in the Ural Mountains evidently destroyed several communities. The disaster was never publicized, and it was left to a dissident scientist who had emigrated to Britain, Zhores Medvedev (no relation to the author), to reveal what had happened. In the Gorbachev period, the scale of the accident began to become apparent; it would have been apparent much sooner had Grigori Medvedev's story "The Reactor Unit" evaded the clutches of the censors. Chelyabinsk may have been the first Soviet nuclear disaster, but it was clearly not the last. Moreover, the industry was allowed to remain aloof from the outside world. The International Atomic Energy Agency (IAEA), the United Nations organization that has played a key (albeit highly controversial) role in investigating the consequences of Chernobyl, was not permitted to inspect a Soviet nuclear plant until 1985, when representatives arrived at the "model" VVER-reactor station at Novovoronezh.

The Soviet program saw the development of two major reactor types. The first, the Soviet prototype, was the RBMK (graphite-moderated reactor) described above. There were three "first-generation" stations, built at Leningrad (St. Petersburg), Kursk, and Chernobyl. Leningrad, as noted, was the original model; Kursk and Chernobyl, located only some 300 miles apart, proceeded at a slower pace. All three reactors were 1,000 megawatts in capacity. The RBMK, which was manufactured in the city of Leningrad, was regarded as the most economical source of nuclear energy of its time. Chernobyl-3 and Chernobyl-4 were constructed as twin reactors with a common building, about 100 meters from reactors No. 1 and No. 2 but linked by a long hallway. At the time of Chernobyl, RBMKs made up the majority of the Soviet Union's civilian nuclear capacity. A new station with a capacity of 1,500 megawatts had recently been constructed at Ignalina in Lithuania, close to the Belarusian border, while construction was well under way on stations at Smolensk and Kostroma (third-generation models).

Several scientists at the Kurchatov Institute of Atomic Energy expressed concern about the safety of the RBMK in the early 1970s. More than thirty design flaws were reportedly uncovered, but none was corrected before the reactor first went into service. At least one official was dismissed for trying to draw attention to these defects. The most disturbing feature about the reactor was its instability at low power—one of the factors behind the Chernobyl disaster. Another was the construction of the control rods, which had been produced shorter in length than stipulated in the original blueprint. The point is that the flaws in the RBMKs were well known to the country's most prominent scientists, but remained a secret that was to emerge only after a tragedy had occurred.

The second type of Soviet reactor, the VVER, was designed in the Atommash factory in Volgodonsk, in southern Russia. This model was considered appropriate for both export and emulation, and was also manufactured in Czechoslovakia. The Soviet nuclear program, which envisaged a similar CMEA (Council for Mutual Economic Assistance) program in those East European countries that faced an energy shortage, anticipated that the VVERs would eventually outpace the RBMKs in capacity. The chief area for expansion within the Soviet Union was the Ukraine, the most industrialized republic and the republic facing the most obvious energy shortage in its machine-tool and metallurgical industries. By the year 2000 nuclear power was to have accounted for 60 percent of the Ukraine's electricity production, along the model of France, a country of comparable size and population. Since the plans for the Ukraine were so ambitious, East European nations were invited to invest in Ukrainian nuclear power plants in return for a fixed portion of the electricity produced. Consequently, Poland sent funds and workers to the Khmelnytsky station in western Ukraine; while both Romania and Bulgaria invested in the southern Ukraine station in Mykolaiv province, on the Bug River.

At the various stations, new towns were constructed for the workers. Like other Soviet industries, nuclear power was labor intensive, and the average size of a town that grew up around a reactor with a capacity of 4,000 megawatts was 40,000 people, including employees and their families, service people, and others. For young people, the idea of working in such an industry, which was publicized as absolutely safe, was not unattractive. The Soviet nuclear power industry—unlike its US counterpart after 1979—was declared to be "accident free." Thus the average age of the population in such towns—Netishyn at Khmelnytsky, Energodar at Zaporozhiye, Pripyat at Chernobyl—was in the mid-twenties. Indeed, the young community appeared at a glance to complement the dynamic new industry that promised to solve the energy problems of the Soviet Union, which no longer would be dependent upon the discovery of new oilfields or the application of modern technology to a coal face.

Major river systems such as the Dnieper (Dnipro) in the Ukraine, already harnessed to produce hydroelectricity, were now linked to the nuclear power industry. Downstream from Chernobyl, the Kiev Reservoir fed the Dnieper, and a new nuclear plant was under construction in Chyhyryn, the historic site of the old Ukrainian Hetman state. Even further south, rising at an extraordinarily rapid rate, was the Zaporozhiye VVER, at which reactors were being brought on line at *annual* intervals. From here it was but a small step for the authorities to envisage a new stage in the development of the atom: power for all of the Soviet Union's major cities. Surely, it was proposed, if the industry is accident free, it should be possible to use nuclear power for other purposes. In theory, the idea was sound enough. A reactor was already supplying heat to a tiny mining community at Bilibino in the Soviet Arctic. Similar technology could be employed in the major cities. Nuclear power and heating stations (acronym ATETs) were well under construction by the spring of 1986. The most advanced ATET was the station at Odessa, followed closely by the Minsk station, located about halfway between the city and its international airport. Stations were also planned for Khar'kiv and Kiev in the Ukraine. The progress of the ATETs reflected official confidence in the industry; by the end of the 1980s all of the ATETs were to be abandoned.

THE CHERNOBYL DISASTER: FIRST DAYS

The Gorbachev regime has often been seen in the West as one that heralded immediate change for the Soviet Union. This is not true. In fact, one could equally well argue that until the end of 1985 there was remarkable continuity among the administrations of the last four Soviet leaders—Brezhnev, Andropov, Chernenko, and Gorbachev. In 1985 Gorbachev could be heard extolling the traditions of Stakhanovism and, on one occasion, even praising Stalin. The antialcohol campaign—the major reform of 1985—ended in abject failure. In this same year the nuclear program was reconfirmed. In the Ukraine, the reactionary party leadership of Volodymyr Shcherbytsky showed no sign of ending. In fact, it was subsequently revealed that Shcherbytsky, a former close aide of Brezhnev, cast the decisive vote at the Politburo ensuring Gorbachev's victory over the Moscow party chief, Viktor Grishin. In Gorbachev's favor were his relative youth, at age fifty-four, and his expressed desire for change, particularly in the sphere of the economy.

The spring of 1986 brought no harbingers of forthcoming catastrophe, with one notable exception. In a remarkable article published in the weekly newspaper of the Ukrainian Writers' Union, *Literaturna Ukraina*, a young female journalist from the city of Pripyat, Lyubov Kovalevska, drew attention to certain shortcomings in the construction of the Chernobyl station, which she described as "an accident waiting to happen." The article was studiously ignored by the industry's authorities, although evidently not by the local party organization, which threatened the author with dismissal. *Literaturna Ukraina* had traditionally been critical of the nuclear industry, and since the 1970s (as its editors today proudly point out), the newspaper had monitored failings and defects at Ukrainian nuclear power stations. The article was therefore perceived by the authorities as but the latest in a succession of protests from the antinuclear element, at that time akin to an irritating flea that could easily be swatted aside. As the May Day holiday approached, the mood in Kiev, as in other cities, was one of anticipation of the long holiday weekend.

The initial media reaction in the West to Chernobyl is well known. Various outlets were frantic to obtain information, statistics, and figures in light of the official silence from the Soviet authorities about what had occurred. There was a "media blitz" but without the information necessary to support such frenetic activity. I recall sitting on a panel at a press conference in New York City on 2 May 1986 and being asked the question that was to be repeated ad nauseam: "How many victims were there at Chernobyl?" When I responded that to the best of my knowledge there was no information available, one frustrated reporter bellowed out of the crowded room, "Give us a number, any number!" The reporters' exasperation was understandable. Yet only today are we finally beginning to ascertain what happened in those first chaotic days after the roof of the fourth reactor building was blown apart by a power surge.

Publicly, the Soviet Union reacted to Chernobyl as it

had reacted to other major accidents in the past. There was no statement from Mikhail Gorbachev until 14 May, almost three weeks after the disaster. TASS issued a terse two-line statement on 28 April, and on the following day this statement was published on the back page of two Kiev newspapers—one of the few occasions on which these newspapers released a major news item before their Moscow counterparts (there were good reasons for this, as will be explained below). On 30 April the Soviet press at last acknowledged the disaster, though not its scale. By early May, however, the Soviet press had taken two new directions. On the one hand, it was reporting on the "battle against the elements" at Chernobyl, as fire crews struggled to quench the graphite fire; on the other hand, it was producing lengthy stories about nuclear accidents at stations in the United States and expressing righteous indignation at what it perceived as the West's irresponsible propaganda campaign about Chernobyl. When Gorbachev appeared on Soviet television, he denounced the "mountain of lies" that had appeared in the Western press.

The Soviet authorities subsequently employed their traditional methods to blanket Chernobyl in an official interpretation that was at best a gross distortion, at worst a dangerous myth that cost the lives of more citizens than were killed outright by the explosion. All information on health statistics became classified. The victim count reached thirty-one and was deliberately halted. It has never risen, despite overwhelming and documented evidence that more than 2,000 people died as a result of Chernobyl, either from the direct effects of radiation or from diseases directly related to high-level exposure, in the first few months after the disaster. A government commission was appointed to deal with the accident's consequences. A variety of government ministries sent teams of workers into the area. And a ten-kilometer (eventually thirty-kilometer) zone around the damaged reactor was designated an evacuation area after two Politburo officials, Egor Ligachev and Nikolai Ryzhkov, visited Chernobyl on 2 May.

Although Chernobyl was no longer a secret, there was very little frankness either about the magnitude of what had occurred or about the continuing cleanup and decontamination operations. As Medvedev points out, the government commission issued orders that the public should be "persuaded" that Chernobyl was just a minor accident. Approximately 135,000 people had been evacuated. In August a Soviet delegation led by Valerii Legasov presented a Soviet account of events to the IAEA in Vienna, a report that received wide praise in the West for its openness. Some contrasted so-called Soviet frankness about Chernobyl with US secrecy about the Three Mile Island accident in Pennsylvania seven years earlier. The report in Vienna laid the blame on a series of "incredible blunders" by Chernobyl operators, who had dismantled seven different safety systems before conducting a wretched experiment, evidently to see whether enough power was still being generated during shutdown to keep safety systems in operation. By December 1986 a concrete shell had been erected over the damaged reactor, a tomb—hence the name "sarcophagus." By November 1986 the first two Chernobyl reactors were back in operation, and in December 1987 the third unit was also returned to the grid system. At this juncture the outsider gained the impression that a great battle had been won. Soviet accounts compared this battle with the victory over

Germany in the Great Patriotic War (World War II).

Grigori Medvedev embarked on a quest to tell his side of the Chernobyl story and to expose the secrecy endemic in a long established system. It was a quest virtually unknown to the outside world. Indeed, by 1987 Mikhail Gorbachev had taken on the mantle of an extraordinary reformer, prepared to transform the country through his policies of *glasnost* and *perestroika*. It was a bold campaign that, from Gorbachev's perspective at least, ended in disaster. Paradoxically, the remarkable changes within the Soviet Union made it even more difficult to ascertain accurate information about the Chernobyl tragedy. The Soviet authorities—whether in nuclear, health, or other industries—had gained credibility as a result of the transformation of their society under Gorbachev. When I began to conduct research on the subject myself, I also found a noticeable rift in the opinions of scientific workers in Kiev and their Moscow counterparts. In Moscow the view was expressed that "radiophobia" was widespread in Kiev, that the population had become neurotic on the subject of Chernobyl. Every illness, no matter how minor, was being blamed on the nuclear accident.

Even in the summer of 1989, when I was sitting at the polished conference table of the Center for Radiation Medicine in Kiev, opposite the scientists who were continuing to monitor the first Chernobyl victims, an air of complacency and almost smug satisfaction prevailed. Outside the room, former Chernobyl fire fighters wandered the corridors, many with huge burn scars across their faces and necks. In one room I spoke with a fire fighter and a former operator. Both were too ill to return to work. The fire fighter appeared listless and disillusioned. I was not permitted to interview the two men alone. And when

I inquired about an outbreak of thyroid tumors whose victims were to be found principally in the northern Ukraine and southern Belarus, I was informed that it was the result of a lack of nitrates in the soil. Here one might pause for a moment to consider: as in the official Chernobyl story, there was some truth in this statement. But it did not tell the whole story. The scientists at the center had found a similar answer to almost every question: yes, there are sicknesses here, but they are not a result of Chernobyl.

The Ministry of Atomic Energy was founded after Chernobyl, in July 1986. As Medvedev notes, this gave rise to some hope that its officials would prove more enlightened than some of their Moscow counterparts. It became clear from the first, however, that the ministry wished to continue and even to expand the nuclear energy program and that all other considerations were secondary to that aim. The ministry called in the IAEA to inspect plants that were suspected of having safety problems, and they were all declared to be satisfactory. But by 1988 the lack of official reaction to Chernobyl had given rise to a groundswell of discontent that introduced a new element into Soviet life: national alienation. In addition to a movement against nuclear power—a genuine grassroots movement, in contrast to the officially sponsored movement to remove nuclear weapons from the earth by the year 2000—there was a feeling in republics such as the Ukraine, Lithuania, and Belarus, that Chernobyl was a disaster that had resulted from a central planning system operated from Moscow (the "center") and that the continuing adverse ramifications of the accident were a consequence of that same center "covering up" the real story. Legasov, the fifty-year-old spokesperson for the

Soviet side in Vienna, was reportedly a bitterly disillusioned man; he committed suicide on 27 April 1988. In Kiev, Yurii Shcherbak prepared a manuscript based on interviews with eyewitnesses to Chernobyl, just as Grigori Medvedev was planning to unveil a quite different version of events. The reality was so disturbing that it was scarcely credible.

CHERNOBYL: THE REAL STORY

The Chernobyl disaster occurred at night on a weekend. The experiment was not one that would have entered the logbooks of any genuine scientists. The plant director, Viktor Bryukhanov, was not present. Nor was the chief engineer, Mykola Fomin. The operator in charge was an electrical engineer without previous experience at a nuclear power plant. Yet the actual cause of the disaster appears to have been the faulty design of the control rods, which set off a chemical reaction when they were inserted into the reactor core at the time of the power surge that was to blow the roof off the reactor. Fortunately, there were only a handful of personnel at the reactor site. Unfortunately, there was no protective clothing for the fire fighters who arrived from Pripyat. The roof of the reactor was made of combustible material, and the fire was therefore more severe than it might otherwise have been. When Bryukhanov arrived, he could not believe that the chunks of graphite scattered around the turbine room were from the reactor core. But he was hardly alone in his ignorance. The failure to follow basic safety procedures caused many deaths that weekend.

In Pripyat, that Saturday morning, the population was

oblivious to the nature of the disaster. Some people, having been awakened in the night by the shudder from the power surge, knew that a mishap had occurred. But there were no official reports being issued, no warnings on the radio. Pripyat went about its normal business: wedding parties, soccer games, fishing trips (including in the cooling pond of the reactor, evidently one of the best places in the area for catching fish). Children wandered off to their Saturday schools in neighborhoods where radiation levels were already dangerously high. This surrealistic life continued for nearly forty hours. Many party officials, who were aware of what had happened, simply fled the city. Months later some of them were declared to be "still on the run," a sad but somehow appropriate reflection of their declining role as a moral force in Soviet life.

And what of the party leaders in Kiev and, ultimately, in Moscow? How many of them were aware of the enormity of the event? There are two answers to these questions. The first is that in all likelihood the nature of the accident was known to hardly anyone. The second is that the magnitude of the catastrophe was known to the party leaders in Kiev and Moscow. The Ukrainian party leaders in Kiev appear to have been almost paralyzed with fear, unable to act without some sort of dictum from Moscow. And the Politburo of the Central Committee of the CPSU quickly silenced any efforts to inform the public about the dangers that they faced. In an interview with the editorial board of *Izvestia*, the government newspaper, in 1987, I was informed that the editors had been ordered not to publish a story about Chernobyl in the 27 April edition. The order—the exact source was not indicated—came "from above." The press had thus been silenced. Pripyat was eventually evacuated, but ostensibly for only three days, and very little else was done to deal with the situation at Chernobyl before the arrival of Ligachev and Ryzhkov.

A wider evacuation was then ordered, but it took place to the west and northwest, so that the evacuees formed a bizarre procession, following almost exactly the path of the radiation cloud. They were herded from one danger spot to another. The reactor continued to throw out radioactive elements at least until 10 May. Although more than 450 different types of radionuclides entered the atmosphere and began to fall wherever the wind took them, the most dangerous initially was iodine-131 (responsible for the growth in thyroid tumors); its half-life, however, is only eight days. A more protracted problem was caused by other elements, especially cesium-137 and strontium-90, with half-lives spanning some three decades. Hotspots of plutonium posed and continue to pose problems that will last, for all intents and purposes, an eternity. By 1 May 1986, as the traditional May Day parade took place in Kiev, the wind had changed direction and an ominous cloud appeared over the city. Kievans were exposed unawares to levels of radiation nearly 100 times the natural background levels.

Even more serious was the impact on Belarus, a small republic of ten million people just north of Chernobyl (the border is about thirteen kilometers from the nuclear station), which had seen little serious reform. Nikolai Slyunkov, the local party leader, played a role in the cleanup work after Chernobyl, but the public at large remained woefully ignorant of its predicament. Some 70 percent of the radioactive fallout landed on Belarusian territory, especially the southern part of the Gomel region. Residents in this region were soon receiving radiation doses that were probably higher than in the evacuated zone, although the exact amounts will never be known. Southern Gomel was a quiet farming community where potatoes were the main crop. If the farmers could not see the radiation and were not informed about it, there was virtually no chance that they would see and heed the newspaper accounts published in Moscow or Minsk. Today there is very little attention paid to these unfortunate early victims of Chernobyl; some two million people live in this most severely irradiated zone.

The decontamination and reconstruction process can be compared to a military operation in terms of its willful sacrifice of human life. The process began with helicopter crews flying kamikaze missions to drop debris into the gaping hole of the reactor core. Some on board took photographs, although none survived to recall the event. "Volunteers" were sent to the reactor roof to throw spadefuls of graphite into the reactor core. Coal miners dug a tunnel beneath the damaged reactor to construct a concrete base in order to prevent the shell from falling into the earth's core (the dreaded "China syndrome"). Military reservists were flown in after the first month, and some were stationed in the zone for a full six months, at emergency levels of radiation exposure—perhaps 75 rems, perhaps more; the geiger counters often did not register such high levels. The Ministry of Defense claimed on one occasion that it kept no records of the reservists who served in the zone in the summer of 1986. And there was little attention paid to basic safety procedures. Mario Dederichs, former Moscow Bureau chief of the German magazine *Stern*, took a photograph shortly after the accident of some of the reservists who, wearing no protective clothing, were nonchalantly eating lunch below the open reactor. In May 1989, on the way to the Chernobyl station, I myself saw such men sitting in ditches, smoking, just below signs that warned, "Danger! Radiation!" Why, I wondered, do they do this? Don't they care about survival?

What happened to these reservists? By 1991, as a result of civilian protests (the mothers of the young men being the most vociferous), they were withdrawn from the thirty-kilometer zone. According to official reports from the headquarters of the cleanup operation in Chernobyl, many have since died of heart attacks. According to their own union, more than 7,000 of their number had died by the spring of 1991. Like much else about this tragedy, the figures are almost impossible to corroborate. Doctors point out that there is no known correlation between increased radiation exposure and heart disease, and there is no official data bank on the number of victims. A list of highly contaminated persons that was maintained in Belarus was mysteriously "stolen." A press release from the Ukrainian Ministry of Health issued in the spring of 1990 noted that malignant and evidently incurable diseases of the skin had developed among the cleanup crews. Some have begun to express their views on paper, and one suspects that before long their stories will be revealed to the world, but even in the post-Soviet Russia, the Ukraine, or Belarus, there are still obstacles in the path of such reminiscences.

Similarly, the correspondence between low-level radiation and increased cancers or leukemias remains a moot issue. In Belarus, the scientific investigation into the consequences of the disaster has indicated that there has been a significant rise in anemias, as well as a notable (though not dramatic) increase in leukemias in the irradiated regions, particularly among boys five to nine years old. The issue is a sensitive one because the relationship between exposure to increased radiation and the illnesses endemic in the Chernobyl zone is far from clear. The situation has hardly been helped by the former Center for Radiation Medicine, whose director, Anatolii Romanenko, made the glib statement in 1987 that after Chernobyl the health of the population in the affected areas was "even better than before." The Belarusian public is so disillusioned with the government's efforts to deal with the tragedy that, according to a poll of April 1992, only 10 percent of the population had faith in the effectiveness of such aid. Conversely, more than 60 percent had confidence in a private charitable trust called Children of Chernobyl.

In short, the Chernobyl tragedy continues and expands. Radionuclides are entering the food chain. Radioactive particles in some cases transform themselves into more toxic substances. The economic crisis is responsible for a lack of nutritious food; the life-style of the population is in general far from healthy. In the Ukraine, in 1991, the mortality rate exceeded the birth rate—an alarming statistic. The key question to be asked is perhaps not whether these dilemmas can be resolved, but whether the old system has been completely overthrown since the abortive putsch of August 1991 and the disintegration of the Soviet Union later that year. Are the new states that are emerging significantly different? Would the establishment of democratic regimes guarantee that the sort of information gathered by such scientists as Grigori Medvedev will reach the public and the outside world?

THE POST-SOVIET WORLD:
SOME CONCLUDING COMMENTS

The initial signs are not encouraging. The old ministries that proved so obstructive in the health and nuclear industries have not been dissolved but, rather, have been taken over by Russia. A certain continuity has thus been maintained. The Yeltsin government, through the mechanism of the Commonwealth of Independent States (CIS), has deemed it essential to have some form of joint control over key industries related to defense or military matters: nuclear weapons, the space program, and the nuclear power industry. Subsequently, the Ukraine began to replace the Moscow-based ministries that had supervised the military and nuclear energy spheres on its territory with ministries of its own. The early post-Soviet months saw a perhaps predictable but nevertheless disturbing deterioration in Ukrainian-Russian relations. The Ukraine appeared to contravene some of the stipulations by which the CIS was founded; Russia clearly violated its November 1990 agreement not to question the violability of present territories when it undertook to renew the status of the Crimean peninsula.

This is hardly the place to deal with this dispute in depth. Suffice it to say that the nuclear power industry has been caught in the middle. In the Ukraine, the staff at nuclear stations have experienced a period of demoralization. The number of minor and even major accidents has risen at a startling rate. The Chernobyl station is a case in point: the second reactor had to be shut down late in 1991, and the timetable for permanent closure of the plant was moved forward from 1995 to 1993. Antinuclear sentiment remains strong, but the new post-Soviet states are all suffering from energy shortages that cannot be offset simply by conserving energy. In Belarus, the president of the Supreme Soviet, Stanislav Shushkevich, informed me in an interview in April 1992 that his government was exploring the possibility of constructing, on the territory of the republic, a new nuclear plant based on Western technology. The leaders of the Ukraine and Russia may be less outspoken, but they must also have considered this possibility as a way out of a temporary (it is hoped) energy impasse.

Thus the wheel seems to have turned, or to be about to turn, a full cycle. Moreover, none of the present governments of the three Slavic states can be described as stable. At the time of writing, Shushkevich's position appears to be the shakiest, but significant opposition has emerged to the Yeltsin regime in Russia and the Ukrainian Popular Movement (*Rukh*) has split over whether to support or oppose the presidency of the former Communist apparatchik Leonid Kravchuk. The potential for failure of these post-Soviet governments is high. All three governments have placed their hopes in economic revival, often—a personal view—based on a very naïve understanding of how a market system operates. The "quick route to capitalism" in its 1992 variety seems to be centered on a drastic reduction of the purchasing power of the long-suffering population, or, put more bluntly, the impoverishment of the majority.

The prominence of former Communists is a signal that at the time of the death of the Soviet Union there was not yet a significant democratic movement or even a formation along the lines of the Polish trade union *Solidarnosc*. The press, often cited as the main harbinger of *glasnost*, has been reduced in effectiveness by twin setbacks: the shortage of paper and the traditional tendency to support the native government. There is hardly a better example of this phenomenon than *Moskovskie novosti* (*Moscow News*), which survived the August 1991 putsch by printing issues outside the country, and earlier, after Chernobyl, became a beacon of light in a dark and noninformative world. The newspaper has since then degenerated into a Russophilic organ, and some of its commentaries would not have been rejected by Stalin in the sort of pan-Russian chauvinism that became endemic at the end of the Second World War. Other newspapers and journals have simply run out of funds and ceased operation. Still others have been taken over by outsiders: friends from the capitalist world who can provide sufficient funds and who have perceived the advantages of an operation run by a highly educated but relatively low-paid staff.

The danger here is that the collapse of the old closed society and its replacement by some form of capitalism will not fundamentally change the morals of the new regimes. Has the lesson that the sort of secrecy that surrounded the Soviet nuclear industry and the consequences of Chernobyl is highly dangerous really been learned? How long will it be before the leadership of the three Slavic states (the main sufferers from Chernobyl) are run by statespersons with no previous affiliation to the former Communist Party? Is there any guarantee that a healthy, democratic press will emerge, or environmental agencies that can monitor dangerous industrial or military complexes and prevent their becoming laws unto themselves? And perhaps the most telling question of all: is the Western capitalist society really a suitable model for the former Soviet regime to follow—that is, can it be stated honestly that many of the problems that led to the

Chernobyl catastrophe are not also present in the West?

It is because these questions have to be raised, because they cannot yet be answered to our satisfaction, that this volume by Grigori Medvedev is so important today. Medvedev portrays a society that is still not familiar to the outside world, a society characterized by bureaucratic obstacles and a lack of interest in revealing the truth about a tragedy. And he portrays an industry—nuclear energy—that could conceal its accidents as easily as spreading a carpet over a hole in a floorboard. Perhaps his most revealing tableau is that of the Soviet censor, ever ready with his pen or the ever-present stamp of authority. Despite his scientific expertise, Medvedev represents everyone who has ever been stymied by this system: the bureaucratic red tape; the bewildering succession of different authorities, all of whom must give their assent before a manuscript can be published. Few will bemoan the death of such a system. Perhaps even fewer can be confident that a new edifice will not arise like a phoenix from the ashes. The Chernobyl tragedy marked the end of an epoch, and it remains a testimony to the corruptness and fallibility of the Soviet system. *No Breathing Room* is an indictment of that same system, but it also carries the message of hope that one person, with courage, fortitude, and persistence, can make a difference.

David R. Marples

PART ONE
Long Before Chernobyl

I ASKED LILIA KHOKHLOVA, who had worked as a literary censor at Glavlit (the Directorate for Literature), to describe the basic methods of Soviet censorship. My question was more than purely academic, because as a writer I was deeply interested in hearing a former censor's own assessment of her work, from the moral or human point of view, and also from that of the protection of state interests, as she understood them.

Since meetings between Soviet writers and censors were absolutely forbidden (in the Soviet Union only editors could meet with censors), I was naturally looking forward to talking with a living incognito and hoped that she would be able to offer an impartial account of the workings of our censorship.

Our conversation took place on 25 June 1990 at a Moscow publishing house where Lilia worked, no longer as a censor, but as a current-affairs editor.

In response to my question—which was posed at the very moment that Soviet censorship was breathing its last, as the new press law had already been passed and was to go into effect on 1 August 1990—Lilia said firmly and with a hint of irritation, "I have signed a written promise not to disclose the working methods of Soviet censorship, and therefore I am unable to answer."

I replied that, having been a writer for ten years, I already knew something about Soviet censorship, and that the censors' actions against me and my works had frequently made me extremely angry. I had felt besieged from all sides, with no apparent way out, despite my best efforts. Nonetheless, a way out had been found.

"I'll tell you how Soviet censorship really works," I said. "They drive the author and his writings into a corner and either destroy him or make sure he never gets into print, isn't that it? An old poet once told me that, sometime back in the forties, a Moscow journal promised to publish his poems and told him to come back in a week. When he went back a week later, the doorman said, 'But there's nobody here; everybody, the entire staff of the publishing house, has been put in jail.' Big Daddy didn't mess around . . . "

"And who are you calling Big Daddy?"

"Stalin, who else?"

"I wouldn't use language like that about Stalin. After all, he was a major figure in Soviet history. And, anyway, there have been many honest people working for the censors, real patriots who care about Soviet culture, just as they care about what happens to good writing of all kinds. It's through people like them that many books by Soviet authors have been published and have lost very little along the way. And your books, too, they were published," Lilia said with a smile. Then, quite abruptly, she asked, "If you get rid of the hangman, does death then disappear also?"

"No, death stays. But it's preferable for it to be from natural causes."

"But it stays, doesn't it?"

"Tell me, what made you go into censorship?" I asked.

"I was born into a squeaky-clean family," Lilia replied. "My parents were Communists whose every thought was devoted to the well-being of the fatherland. When I was offered a job at Glavlit, I thought about it for quite a while before I eventually agreed to take it. I imagined that I would be in the front line of the ideological struggle and would serve the cause of the party."

I asked whether she understood that censorship was, in essence, a huge barbed-wire barricade around the actions of the government and the party, and that the current structure of the Soviet government was like an octopus, each of whose tentacles was a ministry of something or other. In the central government alone there were dozens of such ministries; the barbed-wire coils of censorship protected the space not only around the governmental octopus but also around each ministerial tentacle, and wherever those tentacles touched the ground there arose a zone of death or ecological disaster.

Lilia agreed with my arguments, but she kept pointing out that "even if you execute the hangman, death does not go away," and that nothing could change that.

I asked her to put her thoughts about censorship down on paper.

She refused, but suggested a compromise. "I don't mind if you make a story out of our conversation."

Here I must add that Lilia Khokhlova is a very spirited woman who openly fought the construction of the Severnaya thermal power station in Moscow and took part in demonstrations. To the best of my knowledge, opposition to the plant paid off and construction was halted. I cannot help wondering, however, whether things might not have been different if those decades of censorship had never happened, if society had been able to study and criticize many projects that would later prove hazardous to the environment and to human health. Perhaps we would never have had the problems of Chernobyl, the

Aral Sea, the canals built by Minvodkhoz (the Ministry of Land Reclamation and Water Resources) that recklessly tore up and tunneled through my native soil. Perhaps there would have been fewer nuclear power stations, and the Severnaya thermal power station project, against which even a former censor, Lilia Khokhlova, waged a valiant struggle, might never have progressed beyond the drawing boards.

Yet I think back now to 1988, when the censors were quite literally at my heels, when Glavlit started criminal proceedings against me, when writers I knew told me, with barely concealed malicious glee ("See? I knew the authorities wouldn't put up with any nonsense; they'll flatten any smart-ass who tries anything!"), that a commission had been set up and that the publishing houses and periodicals would be forced to hand over everything I had written for them.

At the time I was hoping to conduct an interview on the subject of censorship with the editors of *Komsomolskaya Pravda*, who had in fact suggested the idea to me in the first place; but when they found out that I was still not a member of the Soviet Writers' Union, they refused. Apparently they had a policy of not granting interviews to nonmembers of the union. When I inquired who had instituted such a ban, they offered no reply but merely looked up at the ceiling. I realized that it was the Central Committee of the Communist Party, and that there was nothing that could be done.

While I still thought the interview was going to take place, and while I was still invigorated by the prospect of striking back at Soviet censorship, I talked with Natalia Solntseva, who had spent ten years working for the censors. At the time she was employed in the chief editor's office at the publishing house Sovietski Pisatel, which was producing my book. Solntseva had read and written a review of my manuscript. Like the other members of the editorial staff, she thought highly of it. There was, however, considerable doubt as to whether the manuscript would be published. When I told Natalia about my scheduled interview, she looked down and said, "I advise you not to tangle with the censors. They're very vindictive."

As someone who really knew her way around Glavlit, Natalia was a valuable member of the chief editor's staff: she was able to give informed advice about the chances that any particular book stood of being published. In time I came to realize that every publishing house and magazine tried hard to secure the services of such an "agent" on their staff.

"Better not tangle with the censors." "They're very vindictive." "They have the power to do it." "Glavlit is only part of it. The main thing is the ministerial censorship, they send black lists to Glavlit. So . . . "

On 10 February 1989, V. A. Boldyrev, the director of Glavlit, told *Izvestia*: "The question of censorship by the ministries is very much to the point. I have already noted that many restrictions on the press, in force until very recently, were of ministerial origin, designed not so much to protect state secrets as to hide the failings of the ministries themselves."

In an interview with *Izvestia* on 27 July 1990, Boldyrev was even more upbeat: "Many people, and not just in the press, identify Glavlit with censorship, with supervision of the contents of published material. Yet for some years now we have not been involved in that kind of censorship at all."

Would that the head of Glavlit were right.

In the following pages I shall attempt to show how the working methods of Soviet censorship affected me personally over many years. And not only me. Today millions of people are suffering the effects of radiation that is one of the consequences of Soviet censorship. Or, to quote the former censor Lilia Khokhlova, "The hangman is dead, but death is still with us."

On 1 August 1990 censorship, the hangman, died. Yet death from radiation is still with us and will continue to pursue and torment people for years to come.

Long before Chernobyl I had started to warn the editors of periodicals and at publishing houses about the approach of a new Hiroshima, and I asked them to publish my stories about the lives and problems of nuclear power workers and the great danger posed by nuclear power stations to people as well as the environment. But . . .

In April 1979 I was introduced to the deputy chief editor of the Sovietskaya Rossiya publishing house, Nikolai Dimchevsky. I gave him four stories, each of which outlined the possibility of a major nuclear disaster. Dimchevsky took the manuscripts with little apparent enthusiasm. Laying his hand on them, he said, "I can't promise anything, because these may turn out not to be literature." But he read the stories quickly, within ten days, and then said, "I am prepared to stand up in any court and confirm that you are a writer." (He's praising them, so he intends to turn them down, I told myself.) "Your stories are professionally done, and the subject is one that everyone needs to know about. Your stories must be printed. When I read them, I realized that a terrible calamity is on the way. They must be printed, but first we need the go-ahead from Gosatom."

I said I would try to get permission, as I worked at Glavatomenergo (the Directorate for Nuclear Energy). I would try to convince the senior people there.

So as not to lose momentum, I gave one of the stories—I think it was "The Operators"—to the deputy director of Glavatomenergo in charge of censorship (at the time I was a department head in the same directorate). After reading the story, the Glavatomenergo censor hesitated for some time, plainly unsure of how precisely to reject it. Eventually he came up with the following answer: "Your story says absolutely nothing about the party. In this country nothing is ever done, nothing ever happens, or ever will happen, without the party."

Despite this setback I did not abandon my attempts to wrest an imprimatur from the bureaucrats so that I could talk openly to my readers.

I first started writing about the subject of nuclear power in the light of my own experience. In 1971 I was treated for radiation sickness in Clinic No. 6 in Moscow. I was really put through the mill; I saw patients in adjoining wards dying from heavy doses of radiation—scenes of terrible agony that I described in *The Truth About Chernobyl* and "The Hot Chamber." For the first time I began to ask myself certain questions: "What are we doing? The fact is that, if worst came to worst, my sufferings and those of the boys dying in the next room would be felt by everybody. That would be the end of life, the end of humanity." For the first time I pondered the moral aspect of working in the nuclear power industry. Until then I had been a loyal advocate of nuclear power stations, taking pride in the great intellectual and engineering achievements they involved. But then I found myself describing deaths that had been caused by nuclear power ten and fifteen years earlier. It really set me wondering. Sooner or later *someone* had to start thinking about these things.

Later, while working in Moscow, I saw how certain powerful people who were quite remote from the nuclear energy field thoughtlessly, as if in a state of incomprehensible euphoria, decided to cover up dozens of accidents at nuclear plants and press ahead with the construction of nuclear power stations in the European part of the USSR. Invisible alarm bells were ringing; the possibility of a major nuclear accident, perhaps a disaster, loomed ever larger. After all, if they covered up accidents at Chelyabinsk (my story "The Reactor Unit"), at Sosnovy Bor, outside Leningrad ("The Syndrome"), and at Tomsk and Krasnoyarsk—in other words, precisely the kind of information that should never have been suppressed—then by some strange, perverse law, there would eventually be a disaster so horrendous that a cover-up would be completely out of the question. In such a case evil and lies would clash head-on with the morality of nature, and it would be morality, rather than people, that would unmask the lies.

It was essential to write, describe, educate, and warn, imploring readers to be vigilant and active in upholding their right to life, health, and accurate information.

Having understood all this, I was not entitled to remain silent—no matter what the price. It was still three years before Brezhnev's departure, six years before April 1985, and seven years before the Chernobyl explosion. It might still have been possible . . .

It was on 26 April 1979 that I gave the manuscripts of "The Reactor Unit," "The Operators," "The Hot Chamber," and "The Expert Opinion" to Nikolai Dimchevsky at Sovietskaya Rossiya. Upon hearing Dimchevsky's initial reaction, I asked him for the official response of the publishing house. After all, he clearly appreciated the stories, yet they had still not reached the printers. Dimchevsky was a writer, the deputy director of Sovietskaya Rossiya; he would not have the last word on the matter.

I soon received a reply. A letter from the publishers praised my stories, underscoring the importance of their subject, pointing out that they opened people's eyes to a wholly new threat to their lives and to the environment, and agreeing that a terrible disaster was probably on the way. However, the letter concluded: "For the last two years the publishing house Sovietskaya Rossiya has switched entirely to the publication of Soviet and foreign classics; for that reason we are unable to publish the stories you have submitted."

Dimchevsky was made uneasy by the rejection. When I told him about the accident at the Leningrad nuclear power station in 1976, during which part of the fuel assemblies had melted and as much as two million curies of radioactivity had been ejected into the atmosphere (my story "The Syndrome"), he became quite upset, saying that a female cousin of his in Leningrad had recently died of leukemia, and that her death might have been related to this radioactive emission. Yet his emotion and indignation did not last beyond a brief emotional outburst. He did not agree to publish my stories in *Sovietskaya Rossiya* and advised me instead to try Inna Borisova at *Novy Mir*.

I went to see Inna, who was senior editor of the prose section of *Novy Mir*, in May 1979. I briefly told her about the nuclear problem and handed her a copy of "The Reactor Unit," which was based, as I have noted, on the accidents at Leningrad in 1976 and Chelyabinsk in 1957 and related events. Inna was a fluffy-haired and pink-cheeked woman who listened to me in silence, focusing her big and practically unblinking blue eyes on me as I spoke. Her apparently sensitive nature reminded me curiously of the patients I had known who were being treated for radiation sickness. Much later I realized that my instincts about Inna were right: she perceived the whole of our horrendous, oppressive reality like x-radiation—a sort of radiation by reality. She felt crushed by the huge weight of life, by the x-rays of ideology. Interestingly enough, it was she who first accepted my story "The Trail of the Inversion."

I should mention that in 1978, even before I had met Dimchevsky, my nuclear stories had been read by Lev Naumenko, a doctor of philosophy who served as the first deputy chief editor of the journal *Kommunist*. In those days I could hardly have guessed that in 1989 this same journal, with a new board of editors, would play a decisive role, together with *Novy Mir*, in getting *The Truth About Chernobyl* published. But these events were ten years away. . . .

After reading my work, Naumenko had said, "You know, it's powerful stuff. It's really a treacherous denunciation of the state in literary form. At a time of nuclear confrontation you have dared to lay bare the failings and negative aspects of our nuclear experience. They'll make mincemeat out of you. People have been hauled off to labor camps for much less."

"Do you think that warning society about a grave danger is a treacherous denunciation?" I had asked.

"But you're attacking the regime."

"Well, fine. To hell with the regime."

"I'm just warning you of the possible consequences."

✹✹
✸

A year later, Borisova read my story quickly. She said that it should definitely be printed and that its publication would have the effect of an exploding nuclear bomb (that is, if it were published in *Novy Mir*). She considered the story important for the protection of human lives and of the environment. Like Dimchevsky, however, she warned that the stamp of approval of the censors in the relevant government departments would be needed. "You need to have the Gosatom stamp on the title page."

Like everyone else, she uttered the word *Gosatom* with a sense of hopelessness. This was because in 1979 government departments such as Gosatom were at the peak of their lethal powers. Soviet troops would soon be entering Afghanistan, and literature, with the exception of Viktor Astafyev, Sergei Zalygin, Vasily Bykov, and a few others, was virtually under the control of ministerial departments.

"I prefer the more relaxed storytelling in 'The Operators,'" Borisova said. "You manage to say quite a lot in a quiet way."

"But 'The Reactor Unit' has a powerful message," I said.

"That's not the point," said Inna, trying to calm me. "'The Reactor Unit' is an exception: it deals with the overall danger to the lives of everyone, including the ruling elite. 'They' will forgive that. And another thing: I understand that Valentin Rasputin was severely beaten near the entrance to his apartment building just recently. He was hit over the head and is now in the hospital. This must have been retaliation by the army or the KGB for his story 'Live and Remember,' and perhaps for other

43

writings as well. They are capable of anything," she said, staring at me with her big blue eyes, as if trying to gauge how well I would stand up to such an ordeal. "They could make mincemeat out of you."

I had heard the same thing from Naumenko.

"Right now what we have to do is publish the story, despite what could happen later," I said.

"There's not a lot I can do," Inna said quietly. "Naturally, I shall give your story to Diana Tevekelyan, the woman in charge of the prose section, but I'm afraid that she's frightfully busy." Borisova's facial expression and tone of voice made apparent that she thought the situation hopeless. She was only too familiar with the journal's methods of operation, and she was judging the chances of "The Reactor Unit" realistically. Clearly the timing was wrong: the events I was writing about would have to happen repeatedly, and then there would either be cover-ups, and continued lies about the absolute safety of Soviet nuclear power stations, or . . .

"It would help if your story had the support of one of the literary big shots, like Yuri Bondarev. We worked together for a time on *Literaturnaya Gazeta*, and I found him very responsive."

"Perhaps you could call him then?" I asked.

"You know, he's a real recluse these days," said Borisova.

"What about Shaginyan?"

"She's really influential and could do a lot. Try getting in touch with her."

By then I already knew who was in charge of censorship in the editorial offices. I was told that it was the first secretary, who was involved not only with censorship but also with the KGB, which had its own literary department, doubtless to keep an eye on intellectuals and assemble files on them. The main key to censorship, however, lay in the safe of the general secretary of the Communist Party's Central Committee. After all, permission to publish *One Day in the Life of Ivan Denisovich* had come directly from Khrushchev's Politburo.

"Another thing," said Borisova, "your story should also have the support of one of the leading nuclear physicists of the day."

Mentally I ran through a list of such people who might support me, but I could think of no one. Sakharov's name came to mind, but he was in disgrace. I never imagined then that he would be the one who would warmly support *The Truth About Chernobyl* in May 1989; in fact, he read the manuscript in one night and within three days had produced a foreword. But that was ten years later. Meanwhile . . .

I got Marietta Shaginyan's phone number from Inna Borisova and immediately called her. It was around this time, in March 1979, that an accident occurred in a bacteriological weapons plant in Sverdlovsk. I first heard about it while listening to a "hostile" radio station. Soon, however, details filtered through from the Urals to Moscow. It seemed that a middle-aged female doctor, a microbiologist, had accidentally (or purposely) broken a flask containing an especially virulent culture of bubonic plague or malignant anthrax that had a peculiarly selective effect. The bacterial cloud thus released had apparently passed out of the building and been scattered by the spring winds. Approximately one hundred people soon died from pulmonary edema diagnosed as "bacterial pneumonia." Although it was difficult to judge the accuracy of the rumors that circulated at the time, news of the disaster quickly spread around the country and left many people horrified. Marietta Shaginyan knew of the accident. I couldn't help thinking at the time that the mere existence of nuclear and bacteriological hazards was monstrous. Yet they were a reality. Given the moral standards of scientists and of our system, it seemed unrealistic to hope for anything better. We have to fight with whatever we've got. And all I had was words.

Shaginyan picked up the receiver. She sounded quite excited. When I got home I recorded our conversation in verse form:

> *I went to see her*
> *   With nuclear material.*
> *She answered.*
> *   And, strange though it may seem,*
> *She said in a splendid young voice:*
> *"I shall soon be ninety-two,*
> *My friend. And I am so tired.*
> *   I really hate you nuclear experts!*
> *Oh, if only I were allowed to see*
> *   With renewed vision*
> *What I see through my soul!*
> *   You are the enemies of mankind!*
> *I cannot hear well . . .*
> *   You have taken apart the cell . . .*
> *And, a further misfortune,*
> *   You have created incurable illnesses.*
> *Write about this,*
> *   The atom is more diligent,*
> *It can be subjected to reckoning.*
> *   But you can no longer stop the wheel.*
> *It will be hard for you to break through;*

> *I wish you well, though.*
> *I shall hang up now.*
> *The sirens of alarm have cut off time."*

At the time I thought Marietta was mistaken. The atom can be subjected to reckoning only on one condition: if the people in charge of nuclear power are highly moral. What kind of morality can there be among such people if every single accident at a nuclear power station, and the irradiation of people and the contamination of water and soil, is concealed beneath a blanket of terrible secrecy, and if any attempt to publicize such matters is regarded as a criminal offense?

Even so, I was convinced that there was only one way to shake people up and make them vigilant—by showing them the whole grim picture of our collective progress toward the nuclear abyss. Seven years before Chernobyl, I was already convinced that there was no other way. Although the system had left people soulless and apathetic, they still wanted to live. Accordingly . . .

Once again I sought advice from Dimchevsky. He seemed moved by my account of my conversation with Marietta Shaginyan, and he suddenly began to describe the old lady's extraordinary daring. "She even scolded Nikita Khrushchev. That really stirred things up. If she wanted to, she could support you. *Kommunist* prints her stuff all the time. In fact, she has access to all the publishing houses and journals." Dimchevsky was so excited his face became flushed; he took visible pleasure in praising someone else's power, especially that of a writer.

"But this censorship—it's a real killer, a monster!" I said angrily. "You know what your Glavlit does? It lays minefields for our future. It tries to wrap up the truth in armor plating. You know, the truth will explode one day, like dynamite. And it's not just truth: it's the truth about the atomic nucleus. And if they try to keep the nuclear calf wrapped up in armor plating, you know what's going to happen. . . ."

"I'm quite sure that sooner or later your 'nuclear calf' will break loose," said Dimchevsky, who obviously liked my unusual choice of imagery.

"Let's just hope it doesn't explode," I said wearily. "Clearly the wave of the future is bioenergy; there's no way of stopping it. There will be trouble . . . "

"Yes," he said, also upset at this prospect. "Nowadays the censors can do what they like, and there's nothing that can be done about it."

Unoptimistically, Dimchevsky picked up the phone and called the chief editor of *Novy Mir*, Sergei Narovchatov. Dimchevsky proceeded to speak frankly and enthusiastically, recalling that he himself was indebted to Narovchatov for recommending him for admission to the Writers' Union, and asking him to receive the writer Grigori Medvedev, with his story "The Reactor Unit." He urged Narovchatov to read the story and to consider publishing it. He said it would bring credit to *Novy Mir*.

Narovchatov told Dimchevsky to send me along, that he would be expecting me. I was extremely grateful. Dimchevsky had set the train in motion—quite a substantial feat. His words of encouragement bolstered my confidence and charged me with ever greater energy.

Together with my manuscripts, I carried about in my file a map of the European part of Russia on which the locations of operating and projected nuclear power stations were indicated in red. Sites where nuclear accidents involving the release of radioactivity into the environment had already occurred were specially marked. By showing this map at the same time as my manuscripts I attempted to appeal to the editor's conscience and sense of civic duty, as well as the natural desire for self-preservation. Unfortunately, most of the editors just stared vacantly at the red network of nuclear power stations girdling Russia and handed back the map without a word, and without looking up. Some of them asked me to leave the map with them as a souvenir, while others promised to use it as a means of steering clear of those places where nuclear power stations were located.

Carrying this very map and the manuscript of "The Reactor Unit," I entered the reception area at the office of the chief editor of *Novy Mir.* The time, I believe, was June 1979.

Through the poetry and short stories of the poet Mikhail Lvov, a friend of Narovchatov's, I knew that Sergei Narovchatov was a towering blue-eyed figure who resembled a hero in Russian legend, that he was a poet and a veteran of the Soviet-Finnish and Second World wars. I expected to find a mature, powerful man, and I was already warmly and gratefully disposed toward him. Against all expectations, here I was, an unknown writer warning about improbable dangers that no one knew anything about. Yet I was driven by some mysterious force, mainly the force of knowledge and experience. I knew that not only the past but also the future was wrapped up in the present. It simply remained for us to reach it. I did not want Chernobyl; nor, I am sure, did anyone else. Unfortunately, the ruling regime seemed to be calmly pursuing a course that led inexorably toward disaster. I did not have the right to stop.

When I got to the second floor, I at once recognized

Narovchatov, who was walking along the corridor from the chief secretary's office to the reception room. He was bald, of medium build, with yellowish tinges in the whites of his dull blue eyes, and he had a generally unhealthy, weary appearance. He glanced at me silently, as if recognizing someone. I introduced myself and followed him into his office.

The office of the chief editor of *Novy Mir* has played a significant role in our literary history through the past few decades. This is where the illustrious Tvardovsky worked—*Novy Mir*'s first, recalcitrant editor, who put every ounce of his energy into the fight for genuine literature. Solzhenitsyn came here frequently: it was in this office that the decision was made to go to press with *One Day in the Life of Ivan Denisovich*, which I read at top speed in 1963 on my way to the nuclear-powered navy base at Zapadnaya Litsa.

Standing behind his desk, Narovchatov had a remote manner; as he looked at me through blurred eyes, he somehow seemed a throwback to the 1940s. His handshake was limp, soft, and dry. It occurred to me that he was very sick and that receiving me might be a strain on his health. I quickly gave him the notebook containing my story; he skimmed through it, gauging its literary merit with an experienced eye. Nervously, perhaps feeling the pressure of time, I unfolded the map of central Russia with the red network of nuclear power stations and explained its purpose.

Narovchatov listened without a word and glanced somewhat hastily at the map. Then, sounding less remote than previously, he said, "I'll pass this on to Diana; get in touch with her."

I thanked him, said good-bye, and returned to the waiting

room. Narovchatov followed me; he seemed more vigorous than before, even a little flushed with excitement as he asked the managing editor, Valentina Ilina, to forward my manuscript to Diana for her comments.

As I later discovered, this was a great honor for me. It was usually much harder for unknown writers to get their works read. I was truly indebted to Narovchatov. Again the freight car had been given a nudge, and this time it had actually started moving.

In those harsh and apathetic times, "seditious" works of fiction were often nudged forward on their way toward publication by a myriad of barely perceptible editorial gestures and impulses on the part of publishing houses and journals. And they did start moving, like ghostly freight cars, often unnoticed and at times almost chaotically; some were eventually derailed by censorship, while others advanced relentlessly, gaining speed and assurance, toward some symbolic, rational junction. There they joined huge formations of culture and literature, and the sirens of the powerful locomotives of Russian literature could be heard summoning them forward.

Even the slightest gesture helpful to a writer was clearly valuable.

Such gestures, however, were the exception to the rule—the rare surviving patches of green grass on the gigantic tectonic plates of editorial ideology, which pressed down relentlessly, crushing and destroying the careers of many gifted writers.

But why all the fuss over my stories? Was it worth fighting, hanging around editorial doors in the vague hope of reaching a distant reader and making my point to someone? Here is a brief account of the stories.

"The Reactor Unit" (1979), which deals with the

theme of impending nuclear disaster through an analysis
of the life, work, and morality of nuclear power experts,
was inspired by events that took place at Chelyabinsk and
Sosnovy Bor (the Leningrad nuclear power station). The
story tells how nuclear power specialists came to take nuclear contamination for granted, and how everyone else
trusted them; it describes the cover-up of accidents, the
official advocacy of nuclear energy at all costs, the nuclear bomb that kills without exploding, the radioactive
contamination of thousands of square kilometers of land;
it also tells how people died, and how news of their death
was kept from the general public.

"The Expert Opinion" (1978) focuses on the expert
analysis of a nuclear power station with an RBMK-type
reactor, the battle over the project, and the project's
moral implications; the story directly anticipates the explosion of the Chernobyl reactor.

"The Operators" (1978), set at night inside an operating nuclear power station, concerns the fragility of nuclear technology, the people who control this technology—their fates, joys, and woes—and the peaceful atom as
a lethal sword of Damocles.

"The Hot Chamber" (1976) takes place in a biophysi-
cal institute, and its central character is a scientist dying
from acute radiation sickness. The story describes the extreme stubbornness of certain "scientists" who have long
lost their moral sense and who are determined at all costs
to prove that peaceful nuclear energy is essential and even
inevitable; it also considers the altruism of certain others
who are anxious to prevent the murder of volunteers for
the sake of illusory advances.

In other words, these stories offered a preview of the
Chernobyl disaster seven to ten years before it actually

happened. Chernobyl was already present within us, having been born inside our own minds, yet for a long time we persisted in heaping praise on our senseless accomplishments. All the while our air, water, food, and land became increasingly contaminated. And we, blinded and deafened, unwilling to grasp the truth, continued to die.

Diana Tevekelyan, a very powerful woman, read "The Reactor Unit" and "The Operators" quite quickly, in less than two weeks. I do not know whether she passed her findings on to Narovchatov, although I suspect that she did; with me, however, she remained extremely formal, tense, and distrustful.

"I see what you are trying to say in 'The Operators.'" (The story contains a scene in which the plant workers listen to various "hostile" foreign radio stations; these hostile "Voices" are now considered friendly, of course—a change that Tevekelyan could hardly have predicted.) "You know, it sounds a bit out of keeping with the times. And then there's the Voices and your laughter. I know you were laughing, don't deny it." I conceded Tevekelyan's point with regard to "The Operators"; next came "The Reactor Unit." "'The Reactor Unit' is shocking," said Diana. "Is that really how things are? It's mind-boggling! What are we coming to?"

"Death, that's what we're coming to," I replied.

"This is terrible!" Tevekelyan exclaimed melodramatically; apparently she was convinced that the opposite was true. "But these are good stories. You certainly know how to write." Here Diana was slipping me what we writers called a "sedative," so that I would calm down and not go nervously from one publisher to the next, offering my red-hot radioactive fiction and generally stirring things up. OK, so someone spilled something somewhere, so what? Why make a fuss about it?

I left empty-handed. Throughout the summer I worked at nuclear construction sites where new reactor units were being prepared for start-up. This was immensely depressing. The government's intentions were quite clear: the nuclear capacity being brought on line was constantly increasing, the target being ten new reactor units each year in the very near future. The glorification of nuclear power, which was now at its height, resembled drug-induced mania. People were abandoning agriculture, deserting their villages, and flocking to the nuclear power station sites in the hope of finding a comfortable, easy life inside the lair of this highly desirable reinforced-concrete nuclear beast. Former tractor drivers were either operating bulldozers and cranes at nuclear construction sites or being retrained to operate such equipment, thus beating their plowshares into nuclear armor. The main motives for this metamorphosis were money and food. Shortly after yet another new nuclear reactor unit was started up in the fall of 1979, I returned to Moscow with a sense of impending doom.

At the joint meetings of the government directorates in charge of the construction and operation of nuclear power stations, Vladimir Marin, head of the Communist Party Central Committee's nuclear power branch, pointed out that whereas the food program was unlikely to be realized any time soon, the power of the peaceful atom, and an increase in the country's nuclear capacity, could be crucial to success, even in the food sector. Each megawatt reactor unit brought on line allowed three million tons of oil, at two thousand roubles a ton, to be released for sale abroad. In turn, such a sale meant wheat, clothes, and lots of other goods.

No one believed in the working and the peasant classes. Indeed, it was hard to expect much from them, as they had been turned into the mythical masters of the Union and of all Russia. And, being the masters, they felt no need to work—let uncle take care of them. The annual turnover in the labor force at nuclear construction sites rose to 57 percent. What kind of quality could one expect, if the intermediate-level worker on the construction projects of the century were no better than third-ranking workers?

There was an obvious need to talk about peaceful nuclear energy, and to show people what lay in store for them. People needed to be aroused, to be stimulated to take an interest in life as well as responsibility for what happened to them. If I and many others succeeded in achieving only 1 percent of what we set out to achieve, or even in arousing only one person, it would still have been worth it.

In the fall of 1979, while preparing once again to find a way to reach my readers, I happened to see on television a writer whose works I had read and greatly admired: Viktor Astafyev. This was the first time I had ever seen Astafyev. I watched eagerly and listened intently as he delivered a truly human message that cut clean through the deadening morass of official utterances. As I watched and listened, it began to dawn on me that if this man had just one chance to help me, he would do it. That very evening, I phoned Astafyev.

I should point out that the "period of stagnation," though approaching its end, was still potent. Soviet troops would shortly be invading Afghanistan, and zinc coffins containing the remains of those killed in combat would soon be returning home. The viselike grip of ideology suppressed all life, and it appeared to most people that Andropov's KGB had everything completely under its control. On close scrutiny, however, even the KGB no longer seemed omnipotent. I knew from my own personal experience that it, too, was floundering in the mud of stagnation. But more about that later.

Around this time Inna Borisova gave the manuscript of my story "The Operators" to Narovchatov's deputy, Mikhail Lvov. This was a bold move, as she was bypassing Diana Tevekelyan, who had virtually rejected the story. Borisova's aim was simple: if Lvov, who was friendly with Narovchatov, accepted it, then Diana could be skipped over.

Lvov did accept the story and wrote a review which, though enthusiastic, was inadequate in that he recommended deleting the scene in which the workers listen to the "hostile" radio stations. Mikhail had no idea that in ten years those same "hostile" radio stations would become friendly, and that the Voices from the Other Side of the Hill would sign agreements on cooperation with our own radio stations.

Lvov failed to bypass Diana Tevekelyan. He just did not have the stomach for a fight with her. Though his review signaled a sincere desire to publish "The Operators," he would not usher it into print. Inna Borisova angrily accused him of being weak, despite his sincerity and honesty. I am inclined to think that this weakness was a lifelong stance of his. This principle—of not getting in the way, of avoiding a fight—helped him stay for a long time in the administrative saddle. Whenever Lvov felt that an encouraging assessment or a kind word would cost him nothing, he felt free to be entirely sincere. The truth, however, is that in those harsh and apathetic days such utterances could prove very costly. I am infinitely grateful to Mikhail Lvov for his kind words about my work.

At *Novy Mir* I got the phone number of Viktor Astafyev and, with no introduction whatsoever, I called him up. I told him briefly who I was and I talked about my stories, the issue of nuclear energy, and my unsuccessful trips to the publishing houses. Instead of hanging up on me, as he might have, Astafyev listened attentively. In reply he told me how *Nash Sovremennik* had recently published, with some difficulty, a story about an irradiated soldier; there had been a fight with the censors that had caused part of the story to be cut, but even so . . .

I asked whether I could send him my manuscript. At the time, I was unaware that, as a result of an injury in the Second World War, Astafyev had lost part of the sight in one eye and found reading difficult. Had I known this, I probably would have been afraid to ask. Despite his bad eyesight, however, Astafyev uncomplainingly tried to read my story and offered to help as much as possible.

In order to save time, Astafyev initially steered me back to *Nash Sovremennik*.

"I'm going to a writers' meeting at Staraya Ruza, just outside Moscow, and then on to Finland. Call me again in about two months. But in the meantime go and see Karlin at *Nash Sovremennik*; like Pushkin, he's called Aleksandr Sergeyevich. Tell him that Astafyev sent you and wants him to read this. If it doesn't work out, call me back."

I did just as he suggested. I went to see Karlin at *Nash Sovremennik*. The mention of the name *Astafyev* worked on him like a magic password. Karlin took the manuscript of "The Reactor Unit" and promised to read it without delay.

We talked. Karlin told me that he had once worked as an adjuster for Minsredmash (the Ministry of Medium Machine Building), at Krasnoyarsk-45, and was thus already familiar with nuclear energy. He spoke vigorously, in a thick bass.

Karlin read the story quickly and was moved by it, especially the scene in the village. "It's all true, that's how it really is," he said. He invited me to visit his office again.

I did. This time Karlin's bass voice was a little less vigorous. He said that I had touched on a terrible subject, one that was subject to the strictest secrecy. "What rivers and seas are you talking about?" he asked. "Not Shevchenko, or Chelyabinsk-40? That was Minsredmash, and the bomb. They make sure no one ever gets anywhere near, so that no one will ever find out. What are you getting yourself into here? And then you have these historical parallels, the comparison between the accidents at Chelyabinsk and the Leningrad nuclear power station. You know, there's no way 'The Reactor Unit' is going to make it through Glavlit, they'll nip it in the bud."

"So you're not going to take it on?" I asked.

"We just wouldn't be able to do it," Karlin replied, blushing deeply.

"Is that what I should tell Astafyev?"

Karlin said nothing. This writer is really slow to get the point! A short while before, Karlin had published *At the Last Boundary*, a novel by Valentin Pikul, in *Nash Sovremennik*, and the dust had not yet settled. . . .

In fact, about six months later Karlin was invited to quit the journal. I found this out quite by chance, when there was something I needed to ask him and I tried to reach him by telephone at *Nash Sovremennik*. I was told curtly, "Karlin no longer works here."

"Where does he work?"

"Karlin works at home," came the unsympathetic reply.

At least he works at home, I thought.

In those days one had the sense that the editorial boards of the literary publishing houses and journals were an integral part or at least an appendage of the ideological arm of the Communist Party's Central Committee. The first deputy chief editors paid almost daily visits to the cultural department of the Central Committee in order to judge the current atmosphere and detect any slight whiff of policy shift. The advancement of each new writer was coordinated with the Central Committee bureaucracy.

At *Novy Mir* I once saw the first deputy editor, Feodosy Vidrashka, proudly announce before leaving for the Central Committee, "You mustn't wear a hat to the Central Committee, at *any* time of the year!"

The abject servility that the staff of journals and publishing houses had developed toward their master—in this case, the cultural department of the Central Committee— was matched by the rudeness and arrogance with which they treated writers.

With some rare exceptions, the formula pervading the whole of the literary world was "Writers are nobodies, editors and censors are everything!"

Once created, a work of literature, even one favorable to the regime, was snatched away from its author and was then warped and adulterated. Each step of the process was viewed as a favor to the author. "We have taken your work, now you had better get going while you're still in one piece. You start complaining and we'll kill you!"

This essentially was how the ideological machine of the Soviet state viewed the creative individual, the intelligentsia, and original work in general. If literary and artistic works could have been obtained by some devious means without actual creative artists, the Communist system would have been overjoyed. I am convinced that, had *perestroika* not started, they would have found some way of managing without writers, artists, and composers. The system did not need creative individuals. This fact alone doomed the system to collapse.

From December 1979 to April 1980 I was assigned to various nuclear power stations under construction. In the evenings I worked on new stories about the lives of nuclear power workers.

The reader must be wondering how the author could continue to work in the nuclear power industry while appearing simultaneously to condemn the nuclear industry in his writings. Could guilt be the explanation, or did the author just fail to grasp the contradictory nature of the situation in which he found himself?

This is a fair question, on the face of it. My answer is that I have never fundamentally objected to nuclear energy. It is an industrial sector that cannot and should not be halted. However, certain basic steps need to be followed: the public must be made fully aware of nuclear power's good and bad points; there is a need for a sharp improvement both in the morality and responsibility of industry personnel and in the technology itself; the idea that nuclear power stations are absolutely safe must be banished forever, and replaced by the notion of acceptable risk; the choice of sites for nuclear power stations in various parts of the country must be radically revised, and those stations situated in densely populated regions must be dismantled. Above all, it is essential to remove the veil of secrecy surrounding the nuclear sector, which has already inflicted such damage on individuals and on the environment. Society will be able to make the right decisions only when it has the fullest and most accurate information possible about the state of affairs in the nuclear industry.*

As soon as I got back to Moscow in April 1980, I called Viktor Astafyev and told him about my meeting with Karlin, his favorable response to my story, and his inability to publish it.

"Send it along," said Viktor, "I'll try to read it."

The next day I sent "The Operators" and "The Reactor Unit" by mail to Vologda. Two weeks later I received a telegram that read: "Expect you here by 15 May. Astafyev."

The fact that my favorite writer had invited me to visit him made me very happy indeed. Having learned many a lesson in life through hardship, I am not a sentimental person. Now, however, I was quite overcome by emotion, as I felt that the telegram from Astafyev was both sincere and a guarantee of success. I sensed in it a miraculous power to deliver the truth and full freedom of expression. I was deeply grateful to Viktor, and I will never forget his kindness.

The telegram arrived on 6 May. On the seventh I arranged to take two days off work; my trip would coincide with the annual Victory Day celebrations commemorating the end of World War II. I then left almost

---

* My article "The Green Movement and Nuclear Power" gives a fuller picture of my position on nuclear power.

immediately—not waiting for 15 May—and I reached Vologda on the eighth.

I was already quite fond of the poetry of Nikolai Rubtsov, who had lived and written in Vologda until his death. Astafyev and Belov also lived there: what a town!

The sky seemed enormous. At first I could not figure out why, but eventually I realized that it was because the buildings in Vologda are not tall enough to block out the sky. There are numerous graceful churches in various parts of the town.

I phoned Astafyev at home. His daughter Irina answered and said that he and Maria were in Sibla, ninety kilometers away; Viktor went there to write in the peace and quiet of the countryside, near a river. Irina told me that she would contact her father and he would return the next day, and she assured me that he would have been home if he had known I was coming. Perhaps I shouldn't have come without warning; in any event, I would wait for him. Irina asked me to call back the next day, 9 May 1980.

After checking in at the local hotel, I deposited my briefcase in my room and took a stroll around town. Astafyev lived here, and the whole place seemed beautiful, unique. The atmosphere was that of a northern town, with suffused sunlight, a cold whitish sky, wide open spaces, and a gentle pensiveness. The churches seemed to reach for the sky. As I walked along the muddy waters of the Vologda River, I came upon more and more churches, all of them light and airy; even those that had fallen into disrepair, their belfries now home to pigeons, were still impressive. The whole town conveyed a definite poetic mood, with a subtly forlorn air.

I bowed before the tomb of Nikolai Rubtsov and then

went to a bookshop and bought a new book of his poems. Opening it, I found:

> *I love my fate,*
> *I flee from disturbance.*
> *Plunging my face into wormwood,*
> *I drink deep like some nocturnal beast.*

The hotel, which had apparently been built in the thirties, was nothing to write home about: no hot water and chipped nickel bedsteads (also from the thirties). I slept soundly, with no dreams, and awoke the next morning in high spirits, full of joyous anticipation at the prospect of meeting the man who had actually understood and offered to help.

So as to seem fresher when meeting Astafyev, and so as not to look out of place on the Victory Day holiday, I washed my hair in the ice cold water; it chilled my brain through and through. I was reminded of my childhood in Petrozavodsk. The sky had been this same color, with a calm and spontaneous quality, as if the light were coming from inside the air. And the water in the washbasin had been similarly cold every morning. It brought to mind the tranquillity of the taiga, the vast reaches of Onezh Lake, the thoughtfulness of the north, wisdom and the arcane knowledge of the secrets of life and death, solitude, melancholy, and the slow passage of time. . . .

When I called Astafyev, Irina answered once again. She cheerfully told me that some of Viktor's former comrades-in-arms were also there, waiting. Irina is no longer alive, but I remember her as a well-dressed young woman with fluffy hair; like her father, she was cordial but slightly shy. I shall always remember her.

"He'll be here soon. He's with Mummy, they're on their way."

I found where Astafyev lived on Leningradskaya Street, a light yellow four-story building which, I later found out, was a cooperative. I went in, located the apartment, and then headed out for another walk around town. In the afternoon I stayed close to Astafyev's apartment building, walking up and down Leningradskaya Street. Two other men were pacing up and down that same stretch of street: one swarthy and heavily built, the other thinner but also fairly big. It occurred to me that they must be Astafyev's comrades-in-arms.

When Astafyev eventually did return home, I missed his arrival. Irina had told me that a Volga had been sent to fetch her parents. Somehow I just didn't notice it. By late afternoon the two men were no longer to be seen. I called the apartment again, whereupon Irina told me to come up, as Astafyev had arrived.

When I rang the doorbell on the third floor I could hear a slight commotion behind the door, and a woman's reproachful voice hurrying someone along.

A smiling but slightly confused Irina opened the door. She stepped aside as Viktor Astafyev, the writer whom I so greatly admired and whom I so ardently wanted to see, advanced toward me, focusing on me with his right eye, through which he could see best. He shook my hand.

"Here, hang up your raincoat over here."

Astafyev's pleasant alto voice was clearly that of an honest, upright man.

We entered the main room, where I found the two men I had seen walking up and down outside.

"Here, let me introduce you," said Astafyev. "This is Grigori Medvedev, a nuclear power expert and a writer.

And this," he said, pointing to the man with the swarthy complexion, "is the head of a metallurgical plant in Kurgan. This is the chairman of the Altaisky collective farm. This is the widow of Viktor Kurochkin, the writer—she has come from Leningrad. And this is Maria, Irina's mother."

The table was already set for dinner, so we sat down and drank a toast to the success of our meeting. I apologized for being late.

Having read Astafyev's *King Fish* shortly before coming to see him, I began asking questions about Akimka. A striking image emerged. Viktor was sitting to my left at the head of the table; to his right was the chairman of the Altaisky collective farm; facing him on the couch were the metallurgist from Kurgan and Kurochkin's widow— Galina I believe her name was—a plump beauty with black hair and white skin. On my right was Maria, my host's wife. At the right end of the table, opposite Viktor, sat Irina with her restless two-year-old son, Bitya, in her arms.

Irina was somewhat ill at ease, struggling to keep her son quiet, and she soon left to put him to bed.

The conversation ranged far and wide. Viktor told us about Akimka: how during a gale on the Gulf of Yenisei he had placed Viktor and his son in a boat and set out across the gulf. The water was fairly calm at first, but then it got rough and Viktor suggested that they turn back. Akimka would have none of it.

"I thought we were going to drown," Viktor recalled. "Once we had reached the other side I asked Akimka whether he had been sure we would make it. 'No, I was not,' came the answer."

That's the kind of man Akimka was.

We talked about Viktor's writings, and about Yermolova's production of his works in the theater. They asked me about nuclear pollution. They said that it was almost to be expected that a nuclear power expert should turn to writing, by way of atonement for the sins of the nuclear power industry as a whole.

"It's not a question of atonement," I said in my defense. "We've got to warn people so that they will realize what calamities and dangers are coming their way."

Viktor rose from the table.

"Let's go and have our talk, Grigori, it's getting late. Otherwise there won't be time." While the guests stayed behind in the dining room, we entered the adjacent office. It was about fourteen square meters, perhaps a little less. On three of the walls there were neat, handsome bookcases filled with books from floor to ceiling; a writing table stood next to the window.

Astafyev sat at this table, and I sat in a rather low armchair facing him. A narrow map of the Krasnoyarsk region hung behind him; it extended from floor to ceiling in the space between the window and the end of the bookcase.

Astafyev took the manuscripts of "The Operators" and "The Reactor Unit" out of the table drawer, which seemed to me a highly secretive and important gesture. The text of the stories was typewritten on both sides of the page and stitched into fascicles with a gray cardboard binding. In Astafyev's hands the manuscripts seemed to be no longer mine; they seemed to have gained in weight.

Viktor placed the fascicles in front of him, patted them with his hands, and leaned toward me.

"I read them straight through," he said earnestly, with a faint smile. He began to leaf through "The Operators."

"We're lucky to have such people in Russia. You make a very good case. We will print this story. I have a friend in Moscow, Aleksandr Mikhailov, the chief editor of *Literaturnaya Uchoba*." Astafyev took a copy of that journal off the table and showed it to me. I liked it; the articles in it seemed wide-ranging and substantial. "If he is a friend of mine, he will print it. I'll write a foreword to the story and a letter to Mikhailov. I'll send it to you in a couple of weeks and you can take them to him. I would take it myself while in Moscow, but they're a long way off, somewhere on Novodmitrovskaya. As for 'The Reactor Unit,' that has to go back in the drawer. We can't get it through right now. Let's wait until things improve. I thought that in the nuclear power industry, of all places, things would be in perfect order. But if it's really such a mess as you say, then we're in for trouble. Russia has been fleeced and befouled, and now covered with nuclear filth. But we've got to fight it. . . . Can I offer you a book?" Viktor smiled.

I asked for *The Last Bow*. It was a great thrill to receive this famous book from Viktor Astafyev himself. (The reader will forgive me for waxing ecstatic.)

He turned around, took *The Last Bow* (published in Kishinyov) from among a pile of books on the shelf, and wrote in it: "To Grigori Medvedev, with best wishes. Let's hope that everyone will live, and let's fight for that. Viktor Astafyev."

I thanked him warmly. He wrapped the book in a newspaper and I put it in my briefcase. He praised Aleksandr Satsky and Ernst Safonov, and he said that I, as a writer, should get to know them. Then he asked me how I went about writing: was it from the beginning or the end? He admitted that he sometimes started at the end or in the middle. The important thing was for the work to be present in one's soul, so that one could feel its heartbeat.

I was flattered and delighted by this show of confidence from such a distinguished writer, and I was encouraged by his sincerity.

I told Astafyev that I used a similar working method. First, I have to have a thorough feel for the work; then I form a picture of it in my head; only later, when the characters have acquired so much flesh and blood that they will not let me alone, do I transfer these images into words—a particularly difficult thing to accomplish.

Viktor mentioned that he was planning to leave Vologda for his native Ovsyanka; the decision was final and he had already started preparing for the move, which he expected to take place in the fall. He inquired about the radiation situation in the Krasnoyarsk region.

"Somewhere there is Post Box No. 9, about a hundred kilometers from Krasnoyarsk. Any idea about the radiation levels around there?"

"Not off the top of my head, Viktor, but I shall find out when I get back to Moscow."

"Please find out and let me know."

I promised to do this as soon as I got back to Moscow, and I told him not to worry.

We returned to the dining room, where the other guests were waiting patiently for Viktor. He switched on the color television, which was showing excerpts from a film about the capture of Berlin. In one scene a Soviet soldier picked up a starving German child, who gobbled down the soup in a mess tin. As he watched, Viktor's eyes filled with tears.

"I can't bear to watch children suffering, I just can't!" He stood up and wiped his eyes.

When someone asked him about his health, he complained about persistent high blood pressure. I had noticed a Riva Rocci blood pressure measurement device on the table near the window, and I offered to take everyone's pressure.

"Do you know how?" Viktor inquired.

"I had to learn. My own blood pressure has often shot up after radiation sickness." I found Viktor and his comrades-in-arms all to have roughly the same pressure: 145–150/90.

"That's still pretty good," said Viktor, apparently reassured.

The two of us again withdrew into the office.

"You know, I've been through the mill. When I was working on the newspaper in Chusovoy, my honest reporting got me into the most terrible trouble. We were hungry and cold. It was in these same years that my daughter got a chill on her kidneys; she is still in pain."

I was full of respect for this brilliant Russian writer who was standing right next to me, and whose support was now opening up a new chapter in my struggle. I felt certain now that it was up to me to place before the court of public opinion the truth about the "peaceful atom" and how it had poisoned life.

"Not long ago I buried my father in the clay here in Vologda," said Viktor. "Think of the powerful Siberians—like Misha Strelnikov and Vilka Lipatov—who rotted away in the midst of the stone world of Moscow. I was offered a job in Moscow, in charge of the journal's prose section, right after I'd completed my higher literature courses. But with my eyesight, piles of manuscripts—forget it! I needed to get back to my roots, Grisha, it's really time now!"

We rejoined the others and sat around the table. Viktor's comrades-in-arms were reminiscing about the war. I turned to Maria and, changing the subject, asked about Nikolai Rubtsov.

"He used to come here every day," she said. "He used to sit at this table, where you are now." Quietly, as if embarrassed, she added, "Most of the time he was drunk."

"How did he die?"

"His girlfriend was drunk and bit him in the throat," said Maria, struggling to hide her grief.

"Good God! That's incredible! Where did he live?"

"His house burned down. There's a vacant lot there now."

I found myself pondering the sad fate of a great Russian poet buffeted by the winds of life. "I was strong, but the wind was stronger, and I could not stop." I remembered how thrilled I had been on first reading his poetry, and how greatly he had been admired by millions of people. He was a true genius.

It was getting late. We had one for the road. We tried to call for a taxi, but, as it was 11:30, we couldn't get one. I had a ticket for the night train from Arkhangelsk to Moscow.

"At your age I used to walk that kind of distance," said Viktor jovially.

Clearly emboldened by the liquor, I embraced everybody.

"He can't do it any other way!" said Maria, laughing, as her husband embraced me. We said good-bye and I left.

**
*

Sometime in June I received a package from Viktor Astafyev containing "The Operators"; a sealed letter to Aleksandr Mikhailov; and a second, unsealed envelope with the foreword and a letter addressed to me.

"I think I've done what I promised," wrote Viktor. I was extremely grateful and determined to justify his support.

I had also kept my promise. As soon as I had got back to Moscow, I had sent Astafyev a map of the Krasnoyarsk region showing the radiation situation in the area of Post Box No. 9 and adjacent territories as well as the prevailing wind patterns. The writer had now returned to his native area, and he needed to be spared any unnecessary worries.

When I visited Aleksandr Mikhailov a few days after making an appointment with him on the phone, he unsealed the letter and read it and the foreword in my presence; he also took a cursory look through "The Operators." He immediately said that he would print the story, probably in two installments, in the last issue for 1980 and the first for 1981.

"I'm going to hand the story and Astafyev's letter to the section that works with young writers at *Literaturnaya Uchoba*. They will read the story and get in touch with you. It might need a little editing."

Mikhailov, who in the "period of stagnation" was destined to be the first to break open the seal of secrecy surrounding nuclear power, greeted me coolly. As I left his office it occurred to me that he was going to need plenty of courage. I only hoped that he would be able to publish the whole story unchanged.

In 1980 the first reactor unit at the Rovenskaya nuclear

power station went into service; at the same time construction of the severely delayed first unit at the Yuzhno-Ukrainskaya plant was being rushed. The first million-megawatt unit at the Novovoronezh nuclear power station had just been started up, as had the second units at Chernobyl and Kursk. Construction on the first units of the Smolensk and Kalinin nuclear power stations was proceeding around the clock. All of this frantic work was being done against a background of reassuring blandishments from scientists who had already sold their souls: "Soviet nuclear power stations are the safest in the world! They're absolutely safe! It's true that accidents and breakdowns, with releases of radioactivity, occur in the capitalist world. But, thank God, there's nothing like that here—never has been!"

The sweet treacle of deceit blocked up all the pores of society, mummifying the living organism of the country in a gigantic cocoon of lies. "Accidents don't happen at nuclear power stations! No one has been killed in Afghanistan!"

Academician Andrei Sakharov had been exiled to Gorky; dissidents were increasingly persecuted. Yet even Andropov was already feeling nervous. After all, he was better placed than anybody to grasp the extent to which the Brezhnev regime had putrefied. No one was working properly; crowds of people were wandering around the shops in the middle of the workday; prostitution, corruption, and embezzlement were all flourishing.

The chief of the KGB had set up gigantic new departments within his purview. The purpose of these departments was to engage in economic, literary, and industrial espionage. In a sense he was right to do so, as he doubtless could see no other way to introduce changes in Soviet society, given the severity of the prevailing stagnation. A massive effort to hunt down, buy, or steal other people's secrets and achievements could compensate to some extent for the losses due to the widespread inability to work and the endemic thievery at all levels. The trouble was that under the Soviet system people had forgotten *how* to work. While Russia was being run by a combination of the party-bureaucratic mafia and dictatorial drunken matrons standing behind their counters in small stores, Andropov himself had been drawn into a vicious circle.

In 1980 the bogeyman of the foreign threat, global confrontation, and the arms race were very much in vogue. A safety valve had been opened—the war in Afghanistan. Stagnation in the economy continued, as did industrial espionage in the West and the total suppression of dissent. What else could the system now propose? *Perestroika?* It was already reaching maturity somewhere in the depths of this tired society, deep inside the party. But a whole generation had to pass before it could emerge.

I could tell that the KGB was finding things difficult and was not fully in control from the activity of KGB Major Viktor Barkov. He worked in the agency's economic department, where he headed the industrial units of Minenergo (the Ministry of Energy); Soyuzatomenergo (the All-Union Department for Nuclear Energy); Soyuzatomenergostroy (the All-Union Department for the Construction of Nuclear Power Stations), where I was in charge of production and assembly; and Soyuztsentratomenergostroy (the All-Union Central Department for the Construction of Nuclear Power Stations). Together these four units built and operated the country's nuclear power stations.

Barkov's working method consisted of going around to

the head offices of each unit, talking with the directors and their deputies, and collecting any stray information that came his way.

The following is a representative sample of the questions Barkov asked and the kind of answers he received:

"Tell me, what is preventing the first unit of the Zaporozhiye nuclear power station from being started up within the time limit set by the government?"

"Equipment and cables have not been delivered on time."

"What about sabotage?"

"There's no sign of it."

"Think carefully."

"There's nothing to think about. What kind of sabotage do you mean? The chief saboteur is Brezhnev, go after him."

By 1980–81 Barkov was letting such answers go unchallenged and unpunished. He often reeked of alcohol. He regularly visited each of the section heads, including me. The man who initially steered him in my direction was Viktor Khlopkov, secretary of the party organization in our unit, who later admitted to me, "Barkov really got on my nerves. What for? I felt like telling him: 'Get lost!' Then I learned that you might be able to put him in his place with some solid arguments as a nuclear expert."

Barkov was a crushing, relentless bore. I quickly lost my temper with him. I insisted that there was no sabotage. To which he replied, "Think carefully now." Then I let him have my arguments.

"Look, five reactor units are scheduled for start-up in 1980. Soviet industry can supply equipment for only one or one and a half of those units. There's the answer to your question right there."

"But who planned it that way?"

"Gosplan"—the State Planning Committee—"the USSR Council of Ministers, the minister . . . "

Silence. Barkov scribbled something on his notepad, then departed on a sour note. I felt sorry for him.

Once when he showed up particularly drunk, I tried to talk him out of carrying on with his inquiries.

"Tell me, Major, how did you come to work for the KGB?"

"Through the Komsomol." (I can't remember whether he said it was through the factory or through some design office where he was the full-time secretary for Komsomol activities.)

"Was that when Andropov set up the new departments?"

"Yes, it was at the same time."

"What kind of new departments do you have, if it's not a secret?"

"Economic, literary, a few others . . . "

"But what is there to do in literature? Isn't censorship enough? Do authors need to be flogged as well?"

"They keep tabs on intellectuals. But that's not my job."

"I suppose you mean dissidents. They're coming down hard on them."

"Yes, dissidents, too. They deal with cultural and malicious dissidence generally."

"It's rumored that Andropov writes poetry."

"So they say."

"And what do you do in your spare time?"

"I like records; I'm building up a collection."

I could already see that he was by vocation a collector. Of course, it's good that he collected records; if only that were all he collected.

"Tell me, Major Barkov, once you have gathered a certain amount of information, do you process it and report your conclusions, or . . . "

"No, we lay out the facts on the director's desk. There are many others like me. We take all this garbage along to the director's desk."

"But it's a mountain of garbage! Who sorts it out and draws any general conclusions from it?"

"That's not my job. The main thing is find out whether there's been any sabotage."

The kind of thinking that provided the main thrust for the KGB's work in the economic field was clear. If progressive forces in the party and in society as a whole were to fail, this was the kind of thinking that might prevail.

The editors of journals and publishing houses were understandably afraid, as they were taking great risks by printing my works. The same was true of Viktor Astafyev. Yet he had taken the risk.

I was expecting a letter from the editors of *Literaturnaya Uchoba*. Although my story "The Operators" told only a small part of the overall truth about the lethal effects of the "peaceful atom," I still wondered whether they would go ahead and publish it.

In early August 1980 I received a letter from Benyamin Teplukhin, who was in charge of relations with young writers at the journal *Literaturnaya Uchoba*.

He wrote to say that Viktor Astafyev had sent them my story "The Operators," along with his foreword and his recommendation that the story be printed. They intended to publish the story in the next issue; however, it needed a little further work. So he invited me to come to the journal's editorial offices, on the eighth floor at Novodmitrovskaya, 5a.

Benyamin Teplukhin was a short man, with the dry wrinkled hands of an editor approaching the age of sixty. As I stood in his office he scrutinized me carefully. There was nobody in the large room next to his office, just tables and telephones. I looked around: there must be a hidden microphone somewhere. Ah well, to hell with them! In a furtive voice Teplukhin began to ask me a host of questions: "When did you start writing? Why did you start writing? What are you trying to say? To whom? For what purpose? How did you find your way to Astafyev? It's just as well that you did find Astafyev, and not Yuri Nagibin. If your story and a foreword had been sent in by Nagibin, that would have been the end of that."

"But why?" I asked indignantly. "He's a fine writer, I like his work a lot. Sometime back he did a favor for Astafyev. He was helpful at the beginning. Why should he get into such trouble?"

"He's not one of ours. He's forever traveling around the States. By the way, it's just as well that you did not name the nuclear power station where your story takes place—they would have done you in."

He's in an aggressive mood, I thought. He was already afraid before we started talking, and now he's looking for something to latch onto. "It's an average power station."

The phone rang in the next room. Teplukhin rushed to pick it up.

"Yes, all right, I get you." He hung up and came back to resume our conversation.

"Tell me, what writers do you regard as your teachers?"

"Tolstoy, Chekhov, Dostoyevsky, Gogol."

"OK. What about Gorky?"

"Not Gorky, although at one time I read him a lot."

The phone rang again in the next room, and once again Teplukhin rushed to answer it.

"Yes, yes. I understand." And back he came.

There's got to be a hidden microphone, I thought.

"So are there a lot of accidents at nuclear power stations—in this country, that is?" Teplukhin inquired.

"A lot, but everything is hushed up," I said confidently, pleased now that Teplukhin had begun to grasp my point. I realized that the question was meant to be provocative.

"Sssssh!" he whispered, pressing his forefinger against his lips. "Not a word to anyone about that! Not under any circumstances. You could spoil everything. But if you do everything properly, we can still publish a whole pile of your books. Oh, and there's another thing: one of your heroes is bad news, he listens to the Voice radios and says nice things about Solzhenitsyn. We're going to switch the good things he says to the positive hero, Metelev. That's all decided."

"Astafyev and I never agreed to that. I'm categorically opposed to it!"

"Oh, all right, we'll look into it."

When we met again a month later, Teplukhin told me that on 20 August 1980 the Central Committee had approved publication of "The Operators." The editorial board had held three heated meetings.

What are they getting so worked up about? I thought. They're obviously scared, but they won't be able to refuse Astafyev. I laughed inwardly. Keep at it, boys!

They did just that. Teplukhin asked me: "Don't you have in mind some major figure in the nuclear field today—you know, someone authoritative whose name would lend weight to the publication of your story?"

"You mean Astafyev is not enough?"

"Look, Astafyev is a writer. As he said in the foreword, he is 'far from technology and technical people.' In other words, he disclaimed responsibility for the technical side of the story. There's a gap there that we've got to fill. We need a big nuclear expert."

"No such thing. There was Sakharov, but he's in exile."

"Sakharov is an enemy. Even mentioning his name is risky. You were lucky to get the support of Astafyev—otherwise you might as well have forgotten about getting into print. Your story is appearing in the first issue for 1981, coinciding with the 26th party congress. That's a big honor, don't you think? I must say, you've been very lucky. *Literaturnaya Uchoba* is an organ of the Komsomol and the Soviet Writers' Union, a serious journal. This will be good publicity for you."

They just love to draw attention to themselves, I thought. Is "publicity" really that important? The truth is all that counts.

***

When I read the printer's proofs of "The Operators" in January 1981, I was taken aback, as the text was unrecognizable. Words had been shifted from one hero to another, without any compunction. To an extent Teplukhin had kept his promise. But what about my style, and the words? It suddenly occurred to me that they had moved words about, shifting the position of adjectives in significant ways. Present tense had been changed to past. Not so fast, Teplukhin!

What could I do? The melody, emphasis, and thrust of the words had been ruined. The text only appeared to be mine. What a lousy thing the KGB had done!

I phoned Astafyev, who recommended leaving the story as it was. The important thing was to get into print. It would later be possible to publish the original text in book form; many writers did this.

I heeded Astafyev's advice and, with the greatest reluctance, I left the text as it was. I corrected a few mistakes and insisted on restoring only a few words in the proofs.

A printer's copy of the first issue of *Literaturnaya Uchoba* arrived in April 1981. Suddenly there was a frantic phone call from Teplukhin.

"We must have the official seal of your department, certifying that there are no secrets. It's touch and go. If you can, come in a Volga sedan—that way you'll look more important. We're trying to intimidate the censors, saying you're a big director."

"What utter nonsense! What kind of seal do you need? Triangular, rectangular, round, embossed?"

"Embossed, of course! Type on the title page of the manuscript 'There are no objections to publication of "The Operators," pages 1–60, a story by Grigori Medvedev,' with the seal and the signature of the head of your directorate. And the secret section should put their stamp on it, too, and fast. We're all over the place . . . "

"What do you mean, you clown? The censors have stepped in just as we were going to print?"

I typed the text on the title page; had it signed by the deputy head of the directorate, Yuri Stepanov; had the secretary apply the round embossed seal; then got the stamp of the first (secret) section and delivered the story to Teplukhin.

"Phew! That's a relief! They almost stopped us!" said Teplukhin. "Where's the Volga?"

"What's the Volga for?" I asked. "Is that in addition to

the seals? There are plenty of Volgas right outside, you can say they're all mine."

Teplukhin was clearly tired of my story or, more likely, of the threat that it posed to him. He had apparently had enough of fiction in general, and my short story in particular.

In June 1981 I received my author's copies of the issue of *Literaturnaya Uchoba* containing "The Operators." I realized that this was only the beginning, and that a hard fight was in store. The important thing was that the word was now out on the death-dealing atom. We would now have to press ahead and present a broader picture.

I was practically glued to the first reactor unit at the Smolensk nuclear power station, where assembly work was being supervised. Then I went to the third reactor unit at Chernobyl, where they were in the middle of the prestart-up period. The pressure-suppression pool and the reinforced watertight compartment had been poorly constructed: the lining did not hold up during tests. They were busy plugging up holes. What was needed was a thorough overhaul; otherwise radioactivity would be released into the soil during operations. I insisted that the defects be eliminated properly, and I reported my findings to those in charge. They were unhappy that I was holding up their work. I recorded this episode in a story called "The Critical Path," which I gave to Arseny Larionov at *Sovietskaya Rossiya*. After first promising to publish it, Larionov then retracted his promise. The Central Committee must have read the story, too, and decided to enjoin publication. Nevertheless, a commission was organized under Academician A. P. Aleksandrov. Start-up of the reactor unit was halted until the defects were eliminated.

In my spare time I was rushing to complete work on

three stories: "A Nuclear Tan," "The Departure from Nucleate Boiling Ratio," and "The Syndrome." A bad accident had taken place in the first reactor unit at the Rovenskaya nuclear power station. Although it had been contained, with no serious consequences, it might easily have been as serious as what happened at Three Mile Island in Pennsylvania. The public had to be told about all accidents. Secrecy had to be ended and public opinion aroused. If the myth of absolute safety at Soviet nuclear power stations was not dispelled and discredited, there would be a disaster. Storm clouds were gathering in the peaceful nuclear sky.

"The Reactor Unit" had to be published. It could help steer public opinion in the direction of common sense with regard to "peaceful" nuclear energy. But already Narovchatov, at *Novy Mir*, had refused to publish the story.

I took "The Reactor Unit" to S. A. Baruzdin, at *Druzhba Narodov*, where it was read first by Inna Sergeyeva, head of the prose section. The story moved her to tears. "Where are we headed? We're going to ruin literature!"

"What about life?" I asked.

"Yes, that, too . . ."

She passed the manuscript on to Baruzdin, who called me a few days later at work—this seemed to be the custom at the time—to congratulate me on an undoubted success. When I asked whether he was going to publish the story, he replied that he would try. But he said so without any particular enthusiasm, the same way he had congratulated me.

A year later Brezhnev was dead, and then we had fifteen months of Andropov. Inna Sergeyeva said vehemently, "It's not so terrible for me, I have no children.

But people like Baruzdin and Ter-Akopian do have children. Why didn't they think about them?"

Inna was wrong, though. As Chernobyl made clear, the atom maims and traumatizes indiscriminately—affecting people of all ages and of either sex, whether or not they have children. In the year before the disaster at Chernobyl Inna Sergeyeva's husband died. He was a talented artist and a remarkable person whom she loved dearly. His ashes were scattered over his native region in Kiev, in the waters of the Dnieper River.

On the anniversary of his death Inna traveled to Kiev, which was not far from Chernobyl. Before leaving, she asked me whether the Dnieper had become radioactive. She was clearly distressed, so I tried to reassure her. The "peaceful atom" spared nobody. It had invaded the lives and spirits of millions of people, crippling and harming them as it went. Back in those days, however, five years before Chernobyl, nobody believed me.

Yet the so-called "peaceful atom" is terrifying not by virtue of its inherent properties. Those who control it are at least as terrifying. In March 1989 at a meeting with readers in Rostov-na-Donu—"The Reactor Unit" having been published in the first issue of *Don* for 1989—an old Bashkir came up to me and asked, "In your story why didn't you tell about the execution of the inhabitants of three villages? After the Chelyabinsk explosion a radioactive cloud spread over dozens of villages. The first three in its path got the highest doses. The people who lived there were doomed. Then someone sent along a battalion of riflemen and everyone was killed, young and old alike, so as to cover up. The victims should have been given treatment, but then everyone would have found out what had happened."

"I didn't know about that. If I had, I would have mentioned it in my story. Do you know of any living witnesses to this atrocity?"

"I was told about it by some people who managed to escape. My own village was not far away. I myself heard shots, bursts of machine gun fire."

"Find some witnesses. What you have told me sounds perfectly plausible, but find some witnesses." I was left thinking that if the point of life is keeping death at bay, then the presence of radiation should make us fight a hundred times harder.

** *

In 1981 "The Reactor Unit" lay in a pile of manuscripts on the desk of the head of the prose section of *Druzhba Narodov*. Meanwhile the clock was ticking relentlessly away.

I also submitted another story, "The Departure from Nucleate Boiling Ratio," to *Druzhba Narodov*. The staff all read it, and once again I was offered a "sedative" by Baruzdin in the form of a congratulatory phone call to my office. Baruzdin confirmed that he wanted to publish "The Reactor Unit," which he felt was the more convincing of the two stories. He instructed me to secure the stamp of the nuclear censors—without which, he assured me, we would get nowhere.

In 1983 or, quite possibly, 1984 I paid a visit to Yupiter Kamenev, the chief engineer at Soyuzatomenergo, who readily saw the merit of my case, applied the stamp, and signed his name.

Yuri Andropov and Mikhail Suslov were both dead. Changes were occurring at the top. I found Baruzdin depressed: he felt that *Druzhba Narodov* had Suslov to thank for all of its outstanding publications, including the novels of Yuri Trifonov. Suslov had understood the importance of publishing such works and had provided valuable assistance.

"With him I resolved the most difficult issues," said Baruzdin. "But I don't know Chernenko, I've never had any contact with him. I'm not really sure he will be able to grasp the significance of the stories we want to publish."

What a mess! I remember thinking at the time.

Ural published my stories. I submitted a collection of stories to the publishing house Sovremennik, and another collection to the Library of the Molodaya Gvardia publishing house, which was also responsible for the journal *Molodaya Gvardia*. Each of these works needed the stamp of the nuclear censors. Once again Yupiter Kamenev came to the rescue.

The editors at the Library of the Molodaya Gvardia questioned the validity of the stamp. Their suspicions were hardly surprising, as I, the author, was able to achieve something that most editors and publishers seemed utterly incapable of. But in their view, since I was a nuclear expert working in the government department dealing with nuclear matters and a member of an expert censorship commission, I should not be believed. Yet they were the ones who had asked me to get the stamp, as they could not. Then, having been handed it, they were fearful. Perhaps they had hoped I would fail. But I had delivered the desirable, contemptible, criminal stamp, an obstacle or a gateway on the way to the truth. Here it is! Now eat it! It sticks in your throat? You think it's a fake? The Chernobyl clock is ticking away: the devil does not sleep.

When the Soviet-Bulgarian journal *Druzhba* published a story of mine entitled "The Core," Glavlit accepted the nuclear stamp of Soyuzatomenergo without question. And certainly Glavlit should have known: they had models of all the censors' stamps and signatures. So there was little actual reason for cowardice on the part of *Molodaya Gvardia*.

The new chief editor of the Library of the Molodaya Gvardia sent the proofs to Gosatom and to Minsredmash, the directorate of the bomb-makers, where they would really tear the book to pieces. But we would endure that, too.

The journal *Druzhba* (not to be confused with *Druzhba Narodov*) has a fine staff of energetic, daring young people: Aleksandr Isayev, first deputy chief editor; and Aleksei Rakov, the very circumspect current affairs editor, who handles all prose for publication in the journal. He is an emotional, highly principled man given to espousing extreme positions. He is an emphatic advocate of a new, reborn Russia. He has three small children. I sometimes feel that he—like all of us, I suppose—is constantly afraid of losing his job. Yet his very real fears about the future of his children and his country are, as they should be, the main driving force behind Rakov's actions.

When I first visited *Druzhba*, both Isayev and Rakov pounced on me. "Let us have a story about young members of the intelligentsia in some research institute, with a little spice in it, some naked women, you know the kind of thing."

"I don't write about naked women; there are more important things."

"Not so!" said Isayev, laughing.

"Yes, there are, Aleksei! The ground is burning under our feet, we're poisoning it. Where are we going to live?"

"Let us have a story!"

So I took them "A Nuclear Tan." In short order I found myself faced with the same questions as elsewhere: Which authors do you like? Which literary figure do you model yourself on?

"Model myself? I don't. I just learn the best I can from the classics: Dostoyevsky, Tolstoy, Chekhov, Gogol . . . "

"What about modern writers?"

"I like Viktor Astafyev and Vasily Bykov."

Rakov reviewed my book of stories for the Library of the Molodaya Gvardia. "Your short stories are good. But you oppose death. We've been told to despise death for the past seventy years. Death does not exist. Since we have had so much of it, it's better not to write about it. You write about murder, about continuing murder by new means. They won't forgive you for that."

At last, a real conversation, some straight talk from an honest man. It would have been nice if I had heard such talk from Narovchatov, the oversize Russian hero; but he must have forgotten how.

"Have you thought this thing out? You know, they think nothing of killing, of squashing people on the sidewalk. They don't kill right away, they torture you first. They'll stomp on your balls, kill your children, rape your wife . . . "

"Who?" I asked.

"What do you mean *who*? The KGB! You're pretty ignorant."

"But there's a lot I do know. Stuff your KGB agents. Nothing will be worse than death. I shall do what I set out to do. This is the time to think not of wife or children, but of Russia. If we save Russia, no one will kill any children or rape any wives."

"You remember that line in Blok's poem about 'We shall not budge when the savage Hun rummages through the pockets of corpses, burns the towns, and drives cattle into the churches'? Although it was written long ago, it's really about us. Be careful! In Russia someone has to pay with his life for every book that makes it into print. That's the way it's always been. The Greek for book is *biblia*, get it? There is one superbook, the great bible, God's book. And you writers, those of you who expose the bare truth are, of course, closer to God than us mortals. Your books should be like little bibles. Christ suffered on the cross, but you modern orators are scrambling to get your goodies at the writers' cafeteria. Ha ha!"

I asked Rakov about the literary section of the KGB.

"What a dumb question! It's so as to stuff you writers into holes in the ground, like gophers, and if necessary to pour gasoline over you and set it alight. Ha ha! Does that scare you? Nowadays, of course, it's not like in Stalin's time, but they can still stomp on your balls."

"I suppose you don't happen to work for that section?"

"Another dumb question! Perhaps I do. But there are plenty of others besides me—people of conviction, who believe in God, the tsar, and the fatherland. And there are godless ones, too—in fact, more of them. They're out and out stool pigeons, loyal servants of the regime. All the scum in society works for those agencies, always has, always will. They just want to eat caviar."

My book got stuck in the Library of the Molodaya Gvardia. No one at Gosatom was saying anything. Nina Fominichna, the Glavlit censor, was already intimidated by the editor, who told her that the stamps from Soyuzatomenergo were doubtless fake and that this writer could not be trusted. "Then ask Soyuzatomenergo yourselves!" I shouted. The publishers clearly felt that we nuclear experts needed a thorough investigation, as we were pretty much a law unto ourselves. Journals and publishers have never been able to get such stamps as long as there has been a Soviet Union, but a writer could. It was understandably suspicious. Nina Fominichna began to probe deeply, picking on every word.

I asked the editor, Sergei Kamenyuga, to arrange for me to meet with the censor.

Kamenyuga must have been feeling pleased with himself, not only because he was obstructing a book by an interloper—Grigori Medvedev, who opposed death (while thousands of our soldiers were dying in Afghanistan), was an interloper as far as *Molodaya Gvardia* was concerned—but also because he was saving his own skin and his own job. "Soviet law says that writers are not supposed to work with censors," said Kamenyuga portentously, his shiny nose sparkling.

"Look, please help. My book's in real trouble. Don't you care? I mean, you are the editor."

"The World Festival of Youth and Students is coming up soon. Imagine that your book was published just now. The Voice radio stations would immediately start reading from it. And it would be seen as treason. OK, I'm prepared to make an exception just this once. . . ."

How sharp can you get, I thought.

Together we went to see Nina Fominichna, the Glavlit censor, on Sushchovsky Val. She was a heavy-set woman; in her peasant-type tunic dress or smock she had a distinctly rustic appearance. She looked uneasily, probingly at the editor, trying to figure out whether he had changed his mind about the book. She had been given to understand that the book should not be published. When she realized he was not going along with her, her censor's voice rang out: "There are so many prohibited terms in your book. For example, the word *safety* is absolutely banned in works intended for the general public; but you have it on every other line. Another one: 'protection of water.' *Protection* is also not allowed under Glavlit list 78-R. And that list, as you know, was drawn up on the strength of circulars from Gosatom and other government departments. That means that you are revealing secrets."

Clearly it was impossible to talk with this crone. I asked her about her children. She did have children, but she felt that it was not for women and children to go poking their noses into nuclear affairs. "There are enough experts already, without us."

Lilia Khokhlova should have been there to see the methods of Soviet censorship in action! She who had signed a pledge not to divulge those methods . . .

*

In the preceding two years there had been one accident after another at Soviet nuclear power stations. There had been a fire at the Armyanskaya plant; a fuel melt in reactor unit No. 1 at Chernobyl, with radioactivity released into the atmosphere; and a fire in the structure surrounding reactor unit No. 1 at Zaporozhiye. At the Yuzhno-Ukrainskaya nuclear power station some cation filters had disintegrated, releasing ion-exchange ash dust into the primary circuit. When this dust melts, it envelopes and carbonizes the fuel assemblies, destroying them and disrupting the heat exchange.

KGB Major Barkov ran around the directorates under his supervision, and, his eyes bulging, probed for answers. "Why did it happen? Sabotage?"

"No! It was stupidity, confusion, they were racing to get things finished."

"Why were they racing?"

"The minister, the Council of Ministers, Gosplan, and the Central Committee all want more reactor units brought on line as soon as possible. Each one of them saves three million tons of oil a year; the oil is shipped to the West in exchange for petrodollars, wheat, clothes, food."

"Then who's to blame?"

"Minister Neporozhny, who else? He makes all the decisions."

"That's not for us to say. Concentrate on sabotage. . . . Things can't be that bad."

"Oh yes they can, Major!"

Barkov hurried back to the KGB and made his own contribution to the garbage already piled up on the desk of the head of the economic section. What did it all mean? What was the point of it all? It would be interesting to find out who was delivering such garbage to the desk of the head of the literary section.

Aleksandr Isayev called me from *Druzhba* and said, "We've read 'A Nuclear Tan' and find it shocking. But let's face it, Grigori, we won't be able to pull this off. If by some miracle this thing gets published, Georgi Markov will hand you a membership card in the Writers' Union on a silver platter."

I did receive a membership card, but not from Georgi Markov; this was 1984, after all. I received the card from Aleksandr Mikhailov, former chief editor of *Literaturnaya Uchoba* and now first secretary of the Moscow Writers' Organization, who had published "The Operators" in 1980. And I received it not in 1984, but in February

1990. I still had an eternity to live, dragging my cross uphill.

Aleksei Rakov shouted at me, "They'll tear you to pieces! You'll never survive!" And then, in an emotional tone, he exclaimed, "But I beg you, restrain yourself. Report the truth, enlighten the stupid, offer some hope by rationalizing the danger."

"Together, Aleksei, together. Let's carry our common cross together right to the end. We should all be shouting out the truth at the top of our voices."

The ticking of the Chernobyl clock sounded louder, closer . . .

*

I phoned N. I. Yeremeyev in the press section of Gosatom. His boss, the section head, Odoyevsky, had a distinguished aristocratic name. Yet how degenerate he had become! He was now the custodian of the truth about death—nuclear death—and for whatever reason was preventing the general public from finding out the truth. I doubted very much that Odoyevsky could ever be budged.

"I'm glad you called," said Yeremeyev. "Why don't you come on over?"

Yeremeyev met me on the street, outside Staromonetny, 23, and did not invite me up; instead we found privacy at an empty table in the security guard's office. Yeremeyev was a vain man, short and bald and ill-shaven, with a shiny red face and shifty half-closed eyes.

"Your book demolishes the image of the peaceful atom, which we have spent almost four decades building up on behalf of the Soviet state. What will foreigners think of us?"

"The truth about the nuclear danger must be told, the danger of the so-called peaceful atom."

"There's no danger," said Yeremeyev. "There is a nuclear missile danger. But since you work in our field, you yourself must acknowledge that the peaceful atom is just that, and it's safe. I must admit that your book is the first of its kind by an industry professional. We follow everything in world literature on the subject. You're the first professional, and that's where the danger lies. We're going to fight you. We keep our finger on the pulse of all literature on nuclear matters, and it's out of the question for a single line to get into print unknown to us."

"Oh no it's not! In fact, it's already happened with 'The Operators.'"

"That was just a freak occurrence. The same mistake won't happen again."

The deadly countdown to Chernobyl continued. Although the momentous event was practically upon us, types like Yeremeyev and Odoyevsky, watchdogs at the gates of hell, went on compounding evil and adding to the catastrophe's diabolical potential. We were now counting in minutes, not hours or years.

Yeremeyev was wrong: he was keeping his finger on the pulse not of time but of the Chernobyl disaster. Use your head! Look into the future! It's no good . . .

Molodaya Gvardia soon received a top secret letter from Gosatom, signed by the director of the scientific and technical division of Minsredmash, Dmitri Dmitriyevich Sokolov, known as Dim Dimych for short. The letter said, "Grigori Medvedev has discredited the Soviet nuclear power industry. We do not need writers like him. What we need are writers capable of rousing the population to great feats in the name of the great ideas of

Communism. The nuclear accomplishments of our people are also a significant contribution to the building of a bright future."

Dim Dimych, I thought, you obviously don't know for whom the bell tolls. It also tolls for thee.

**
*

Having read Dim Dimych's letter, Aleksei Rakov jubilantly proclaimed, "Hey, Grigori! The Soviet authorities are smart! You're smart, too, but they're even smarter! They could make mincemeat out of you!"

"There's no such thing as 'the Soviet authorities.' There's just the party and the ministerial departments, those are the true authorities."

"We don't need you to tell us that!" The poetry of suppression and prohibition had clearly gone to Rakov's head. I handed him the manuscript of "The Expert Opinion," a story I had written in 1978.

**
*

In the two preceding years there had been changes at *Novy Mir*: Sergei Narovchatov had died. After he had buried his friend, Mikhail Lvov became even more hunched over and started wearing on his lapel the medal given to laureates of some prize or other. The chief editor of the journal was now Vladimir Karpov, a hero of the Soviet Union.

The profoundly ill Chernenko, a latter-day Tsar Fyodor, sounded really inarticulate on television, where he appeared propped up by V. Grishin and D. Ustinov. Chernenko was not the only one short of breath: the entire country, which for seventy years had been taught to take notice of the regularity of the breathing and gait of its leaders, was now also gasping for air.

I collected "The Reactor Unit" from the offices of *Druzhba Narodov*, as Baruzdin made decisions only with the help of Suslov, who was now dead. Apparently he had no contact with Chernenko. Someone who had established such contact was Vladimir Karpov.

When I first went to visit Karpov, he was still in a good mood, having just been promoted to the post of chief editor. He had recently completed *The Commander* and it had been published in *Novy Mir*. Displaying a paternal manner, Karpov accepted "The Reactor Unit" and "A Nuclear Tan" and promised to read them within a week.

Somewhere within the depths of society, in the nether regions of the party, the winds of *perestroika* were slowly gathering strength, awaiting their cue. Tremors from far underground were already being felt. Still muted but ever more insistent, the Chernobyl clock ticked on. The dying system was preparing to inflict one more foul deed on its people.

Two days later Karpov called me at work and invited me to his office.

"Seems strange, but I read your stuff in two days," he said with the magnanimity of a master, looking at me with genuine anguish and apparently expecting an outpouring of gratitude.

Here I should point out that some three months before this meeting with Karpov I had phoned Vasily Bykov and asked him to support two of my stories, "A Nuclear Tan" and "The Syndrome." He agreed to read them, and I sent him the manuscripts. In his reply, neatly written on glossy white paper, which I received a couple of months later, Bykov voiced restrained delight with both stories and described their author as a talented writer. This latter point, I realized, was important less for enhancing my self-esteem than for swaying editorial opinion. The thrust of the letter was such that it could definitely be used as a recommendation for publication.

But Vasily Bykov went even further. Soon afterward, in an article entitled "The Right to a Name" in *Literaturnaya Gazeta*, he mentioned my stories and said he considered it outrageous that they had not yet been published.

I am profoundly grateful to Vasily Bykov for finding the time and the strength to reach out to me at this especially difficult time. It was through his intercession that "A Nuclear Tan" was eventually published in *Ural*.

I showed Bykov's letter to Karpov in the hope that it would bolster his courage. He read the letter in silence and then quietly, with a slight smile, handed it back.

"Yes, I've read your stories. Believe me, I've had a great amount of editorial experience and I can see you have talent. I can say that about only one in every thousand writers who come in here."

Here we go, I thought, another sedative! They're very good at this.

"You have raised an important moral question. There's a strong need for morality today in the scientific and technical spheres. I was particularly impressed by 'A Nuclear Tan.' The heroes of your stories are vivid and lifelike. I've even dreamed about them. But if I put 'A Nuclear Tan' in our journal, I can guarantee you that the censors will take it out. And they'll also send a personal ruling on your case. Now, if you were to alter your workers so as to make them true leaders, members of the party, perhaps only three out of the four . . . Otherwise the censors are going to give you a very hard time. One of your characters decontaminates his penis. Why put that in there? Let's leave out the penis."

"But a married man cannot return home to his wife with a radioactive penis," I countered.

"'The Reactor Unit' has a clear message. The nuclear mess in this country —and not just nuclear—is perfectly evident. But the story also shows that its author has a moral commitment to improve things, and that is very important. This is a story in the socialist spirit."

"Surely it's in a more generally human spirit: for life, and against death."

Karpov handed back "A Nuclear Tan." He said he would keep "The Reactor Unit" for the time being and pass it on to Anatoly Zhukov, a native of the Volga region, who had recently succeeded Diana Tevekelyan as head of *Novy Mir*'s prose section.

I liked Zhukov a lot. He quickly supported my story. He said that it was extremely important and that Karpov should take it straight to Chernenko, bypassing any Glavlits that might get in the way. The story had to be published. This was the normal, honest reaction of a man with plenty of common sense.

The staff of the Sovremennik publishing house had already read and accepted my book of stories *The Operators*. The chief editor, Oleg Finko, called me in for a chat. Now that the book was close to the stage of actual production, he was all the more keenly aware of his responsibility, and the risks he was taking, as editor. A number of people, including Finko, raised the question of the KGB, and the possibility of reprisals not just against me, the author, but also against the publishers. He reminded me that "they" didn't pull their punches, that they would torture me and stomp on my balls.

Again the reference to balls. These publishers seemed to have a fixation. That's one part of the male anatomy that should be left for women to handle, not the KGB.

Rakov was amused when I told him about my meeting with Oleg Finko. "But you didn't know? There really are women working for the KGB who specialize in that sort of thing. A beautiful young SS type comes along with stiletto heels and spears your balls, calling you all kinds of nasty names and threatening to cut them off if you write bad things about your country."

This notion is very interesting. As the term *balls* seems so often mentioned in the editorial offices of journals and publishing houses—invariably in the context of being under the imaginary heel of the KGB—one wonders whether such things actually still happen or whether the fear is simply a leftover from Stalinist times.

*

Andrei Yefimov, my editor at Sovremennik, who is generally a circumspect fellow, came right out and said, "We'll do the book together. I trust you entirely. To avoid trouble, let's have a review done by a physicist or scientist."

At least he's not insisting on a nuclear expert, I thought.

"If I ask for it, they might not cooperate. And I want very much to publish this book. Of course, we'll write an official publisher's request to your scientist."

There are decent people left in Russia, after all! I exclaimed inwardly.

"And the nuclear censors' stamp . . . As a nuclear expert and a writer working in the nuclear power industry, you're a very special case. We've simply got to make use of that. The book will take two or three years to produce, during which time the nuclear censorship stamps will have to be renewed. I shall try to prepare our censor. I'll let her have the manuscript ahead of time, so that she can be familiar with it."

I was delighted that Yefimov was being so helpful, but two or three years seemed like an awfully long time. The Chernobyl clock was ticking away; 1985 was just around the corner.

**
*

I saw Yupiter Kamenev, chief engineer at Soyuzatomenergo, several times before he entered the hospital for a lengthy stay. He complained of fatigue and said that it was becoming increasingly difficult to run nuclear power stations: there was one accident after another, and everything was creaking at the seams. Kamenev believed that nuclear power stations should be handed over to the public utilities. The trouble was that these utilities were not familiar with reactors or with the details of nuclear technology. He hoped there would not be a major nuclear accident, but sensed that one was imminent.

The second man entitled to issue the nuclear censorship stamp at Soyuzatomenergo was Yevgeny Ignatenko. I had worked with him for several years and knew him well. When I showed him the manuscript of my book, he affixed the stamp and wrote a review.

He also had a warning for me. "If the manuscript ends up at Gosatom, watch out. They'll certainly jump all over it. Our stamp and theirs are equally valid. It's just that Gosatom has been around longer than us, and Glavlit remembers them better. Take care, and the best of luck."

I was grateful to Ignatenko. Chernobyl, now only a short time off, would hit him particularly hard. In the meantime he would perform one more noble deed: giving the censors' stamp and writing a foreword for "The Expert Opinion." Ignatenko certainly did his bit. Regrettably, though, we were too late; the system made sure of that. Chernobyl exploded before we got into print.

*

An excited Rakov called me in April 1985, exactly one year before Chernobyl.

"We've read your story 'The Expert Opinion.' It's just what's needed! Your prediction of an explosion is devastating, it really gets to you. There's a lot of documentary material in there, but the story is very well written. The moral imperative is very strong. We're going to publish it. Get us a review from a scientist and the nuclear censorship stamp."

"You'll press hard for it?" I asked.

"Yes, we'll pull out all the stops. Lesha Isayev has known Slezko, who's now head of Central Committee propaganda, since they were in Tomsk together. He'll have him read it. None of us wants a reactor to blow up. But that's not all. This story has opened up the whole of our social abscess. We're going to print it, mark my word!"

I took the manuscript of "The Expert Opinion" to Yevgeny Ignatenko, who read it in my presence using a speed-reading technique. He wrote a brief foreword and placed the nuclear censorship stamp on the title page of the manuscript. I thanked him cordially. As I left, I noticed a pencil portrait of Mikhail S. Gorbachev hanging on the wall.

Over the previous three years, *Ural* had managed, with the support of Viktor Astafyev, to publish some of my nuclear short stories. Valentina Artyushina, the head of *Ural*'s prose section, also insisted on the nuclear censorship stamps, which Kamenev and Ignatenko invariably provided. Even before Chernobyl these two men, by granting permission for the publication of my works, helped to tell the truth about the dangers of nuclear power. May God give them health and happiness.

By the end of 1985 I had sent "A Nuclear Tan" to Valentina Artyushina at *Ural*. Her reply was prompt and enthusiastic: the story was very disturbing and it had to be published. She was sure, however, that the censors would not sanction it, even with authorization stamps, as the issues that the story raised were too sensitive.

I felt like pointing this out to V. A. Boldyrev, director of Glavlit, as an example of "noninterference" with the content of publications. What happened to "A Nuclear Tan" constituted outright meddling with content.

"Well, try and publish it anyway," I said.

But the staff of *Ural* did not have the courage, so I asked for the story back. That was on 10 March 1986. In a phone conversation Valentina Artyushina told me, "I've sent you the story, but it pains me to do so. It should be published."

*

About that time I went to see Boris Yaroslavtsev on business relating to Soyuzglavzagranatomenergo (the All-Union Directorate for Foreign Nuclear Power). The two of us had spent several years working on Minsredmash reactors at Melekess. Yaroslavtsev was very critical of my behavior.

"Listen, Grigori, I was recently over at the Bolshoi Dom* where I talked with Sokolov. He was cursing to beat the band. 'That young writer Medvedev, who has been writing such garbage about us, might find himself taken someplace far, far away, so that by the time he comes back, if he ever does, he will be bald and gray, and not at all young.'"

"Gee, that's terrible!" I said. "They all dream about the days when they could do what they liked and get away with it."

"No, Grigori, you've got to understand," Yaroslavtsev replied, somewhat more mildly. "The principle of secrecy is based on the division of information into separate chunks. If you put all the pieces together in a pile, you get a picture that anyone, including an enemy, can understand. And you've gone and put everything together in a pile."

"That's the whole point. I don't know about enemies, but I think our own poor Russian people, who know nothing about nuclear matters, need to be enlightened. Otherwise we might end up with a Russian Hiroshima."

"There's no need to get angry," said Yaroslavtsev. "I stood up for you. I told them that you, like them, were doing your utmost to improve safety at nuclear power stations. But they're very nasty. It's hard to give up a monopoly on information. You've got to understand them, too."

"I do. And have for a long time. But without much sympathy."

---

* A term that people in the nuclear power industry used to designate the Minsredmash building on Ordynka, as opposed to the Maly Dom, headquarters of Gosatom, at Staromonetny, 26.

The proofs of "The Expert Opinion" arrived in December 1985. The people at *Druzhba* were very proud. If anything bad happened now, they would have been the first to warn of it. Moreover, as the publishers were fond of repeating, the story contains a very powerful moral imperative, and "The Expert Opinion" would stick like a bone in the throats of the enemies of Russia. I should perhaps mention that the journal *Druzhba*, under Vladimir Firsov, is strongly Slavophile in orientation. The point about the mythical "enemies" thus becomes clear. As far as I was concerned, the enemy was the looming nuclear catastrophe. Readers had to be warned, and quickly.

Since *Druzhba* is a Soviet-Bulgarian journal, the proofs were also sent to Sofia for translation. There, by a perverse twist of fate, the proofs were passed to the section head, who happened to be the wife of the chief of staff of the Bulgarian People's Army. This woman promptly told her husband that what she had read was anti-Soviet material. He then read the proofs and reported on them to Dzurov, the Minister of Defense. Dzurov reported to Zhivkov. From Zhivkov word was sent back to Moscow, to the Central Committee, the Defense Ministry, the army high command, and the KGB.

All of these organizations demanded and got the text of the story from *Druzhba*. Two weeks later came the response: "The story is in the spirit of *perestroika* and aims to improve things in the country. Print it."

The proofs were immediately sent to the printers in Kiev. However, these delays caused the story to appear not in the first issue for 1986, but in the second (March–April 1986). By then it was too late. Chernobyl had exploded.

I now felt obliged to write, as well and as accurately as I could, about Chernobyl. Not having been there on the night of 26 April, I set out to collect material, documents, and the testimony of eyewitnesses and participants in the events. On the basis of my experiences as a nuclear power expert and as a writer, by means of documentary and creative research, I began to reconstruct a picture of the Chernobyl tragedy.

PART TWO
After Chernobyl

As soon as I returned from Chernobyl on 16 May 1986,* I called *Druzhba*. The editors were in a state of panic. They urged me to come quickly, as "The Expert Opinion" was being withdrawn from the second issue. When I reached the editorial offices on the twelfth floor at Novodmitrovskaya, 5a, the staff jumped all over me, like hungry dogs pouncing on a bone.

We assembled in Aleksandr Isayev's office, where I told the *Druzhba* staff what I had seen and explained how I understood the situation. *The Truth About Chernobyl* was only beginning to take shape in my mind, although from 28 April I had kept a continuous diary.

So what had happened to "The Expert Opinion"?

"It's like this," said Aleksandr Isayev. "We got a phone call from Ligachev's office demanding that we halt publication of the story. They wanted to know how the author knew there was going to be an explosion. We explained that the author worked in the nuclear power industry and had been involved in providing an expert opinion on the project back in the seventies. But all they could say was to stop publication until the conclusions of the government commission had been published. How about that! They're stealing our victory, our prophecy and warning!"

"Your warning came a bit late," I remarked.

* My trip is described in detail in *The Truth About Chernobyl*.

"What do you mean? It was right on time! The story was written ten years before the explosion. We're such idiots!"

**⁎⁎
⁎**

In response to a call from Andrei Yefimov at the Sovremennik publishing house I stopped by and talked with the assembled staff. I showed the plans for the reactor unit and the prevailing wind patterns carrying the Chernobyl cloud. Everyone wanted to know how safe they would be in these sinister circumstances. I told them what I knew.

The manuscript of my book was ready to go into production. Just to be on the safe side, Yefimov asked for yet one more authorization stamp from the nuclear censors—a "fresh" one, as he put it. I figured out where to get the stamp. Ignatenko and Kamenev were in Chernobyl; or, rather, Kamenev no longer worked for Soyuzatomenergo, having been replaced by Prushinsky. I knew Prushinsky from the Smolensk nuclear power station, where we had worked together on the start-up of reactor unit No. 1.

I visited Prushinsky and explained that the works included in the collection had already been published in journals. I showed him some of the stories. He carefully scrutinized the list on the title page, then applied the authorization stamp and signed his name.

"What's the point of the stamp if the stories are already in print?" he asked.

"The publishers and Glavlit feel that this collection is qualitatively new; that's why we need approval from you."

"OK. And the best of luck!"

I passed the title page with the stamp to Yefimov. The book was already in production, so I wondered whether it might not be a bit late. It was not, however, as the true enlightenment of the population on nuclear matters was only just dawning.

When I called Artyushina at *Ural*, she was practically in tears.

"I feel terrible! We're to blame for not printing 'A Nuclear Tan.' Send it along right away!"

I did as she asked, thinking, Isn't it strange how people believe in the power of the word? But, of course, the Gospel according to St. John! It says that in the beginning was the Word, and the Word was God. And why have we forgotten God? Without God in our souls, with no rudder and no sail, we are drifting soullessly in time and space. Did we deserve such a calamity? Did we bring it on ourselves? Or did we just miss our opportunity? We had a chance to act, but we did not. A timely, emphatic word about the danger would have saved us. But we did not want it; we clung to our official positions, our cozy ways, our lousy creature comforts.

I called Dimchevsky at *Sovietskaya Rossiya*. His voice was breaking with emotion. "We waited until we got bitten by the nuclear monster—bitten hard! This thing has opened up a nuclear volcano!" How dumb he had been not to publish "The Reactor Unit" and "The Expert Opinion." The public should have been awakened to the danger years ago.

True, but now beside the point, I thought. "Are you going to publish my stuff in *Sovietskaya Rossiya*?" I asked.

"But we've switched to publishing only the classics," Dimchevsky replied in a more subdued voice.

And that was that.

Vladimir Karpov from *Novy Mir* called and invited me over. When I arrived, I found the entire staff assembled in the cafeteria. I described the situation at Chernobyl for them.

"That's shocking!" said Mikhail Lvov, in tears. He asked me into his office.

"We're having our own literary Chernobyl!"

"Actually, we have a Chernobyl state. In fact, we now have a Chernobyl planet."

Karpov agreed to publish "The Reactor Unit." Anatoly Zhukov now felt inspired and announced that, although the timing was regrettably late, *Novy Mir* would now publish the story.

Soon afterward a writers' congress convened. Georgi Markov fainted while the report was being read. Karpov seized what he called the "fallen banner" of literature in the "period of stagnation" and was unanimously elected first secretary of the board of directors of the Soviet Writers' Union.

When I phoned Karpov at the Writers' Union, I was told that he was not there, or that he could not come to the phone. I then managed to secure his Kremlin phone number. On the other end of the line he sounded cautious and respectful, as he clearly thought the call was from the Central Committee. On learning that it was just me, his voice hardened.

"Vladimir," I said, "where is the manuscript of 'The Reactor Unit'?"

"It's at my dacha."

"I need it."

"It's a long way out there."

"The story has been accepted at another journal," I said, untruthfully, so as to have it returned all the faster.

Karpov also lied, saying that he no longer needed the manuscript anyway and that he had no further commitments to me. "OK, I'll send a car out to get it. Call back in an hour."

When I called back an hour later on his ordinary telephone, Karpov's secretary answered.

"Have they brought in the manuscript of 'The Reactor Unit'?" I asked her.

"Was that the one Volodya Bogomolov had? In a red folder? Yes, it's here." She gave me Bogomolov's phone number.

So that's it! I thought. Bogomolov had it. Karpov has been handing it around for people to read. That's all well and good, but this circle is a bit small. I'd like the public to read it.

I called Bogomolov and asked whether he had read "The Reactor Unit." He replied hoarsely and with a hint of indignation that he certainly had, and that he liked it a lot. Such praise from a master was most pleasing. I had read his excellent novel *In August 1944* three times.

As Bogomolov went on, he began to swear and curse. "Chernobyl is nuclear war against our own people! 'The Reactor Unit' has got to be published. Only perhaps you could drop the bit about the black chimney stack?"

I thanked Bogomolov and said good-bye. So the old counterintelligence operative was also afraid of the black nuclear chimney stack, intimidated by censorship and ideology, and horribly correct.

Before the start of the writers' congress Karpov had revealed to me the reason for his fear. "I was talking to someone high up." He did not say who; this was already after Chernobyl. "They said that if we publish 'The Reactor Unit' the Soviet Union will be dragged before international courts."

That's really smart! Once again they were suggesting

that I was denouncing the state. It reminded me of the warning I had been given by the doctor of philosophy Naumenko: "'The Reactor Unit' is a treacherous denunciation of the state. They'll make mincemeat of you." What state were they talking about? Or were *they* the state? How could these halfwits still argue that my warning was a denunciation, especially after Chernobyl? They had taken root in their comfortable jobs with all their privileges, and now it was time for a shake-up.

**

In early April 1986, as if anticipating the imminent horror, I sent three short stories to *Neva*. In May Boris Nikolsky wrote me saying that he had found the stories deeply disturbing. *Neva* had decided to put them in the November 1986 issue. (Why November?) Nikolsky wanted me to send the nuclear censorship stamps right away, preferably on official letterhead.

People were already dying from the Chernobyl explosion; radionuclides had settled over millions of wholly innocent residents of both town and country—and the publishers still needed official stamps. I sent what they requested. To hell with them. Nothing had changed: the watchdogs at the gates of hell still bared their fangs. A while later I had a call from Lourie, head of the prose section at *Neva*, who explained that the Leningrad censors were the stupidest and most cowardly in the world, which was why the stamps had to be renewed every two months.

Well, I thought, they reviewed it. But I renewed the stamps.

Around the time the stories were printed in the November issue of *Neva*, I traveled to Leningrad to meet my editor, A. M. Bolshakov. He told me that the censor who had worked on my stories was the wife of a nuclear expert from the Leningrad nuclear power station. This woman had read the stories and passed them on to her husband. Together they decided that the stories had to be printed. Shortly after signing the authorization to print, this woman had been dismissed from the Leningrad Glavlit. She was a brave woman! I do not know her name, but I am deeply indebted to her.

Lilia Khokhlova, had she known of this incident, might have realized that the central issue was not authorization stamps but a person's conscience.

In early June 1986 I spent two hours talking about Chernobyl to a standing-room-only audience at the Sovietski Pisatel publishing house. It was on this occasion that I really first began to address the question of the basic causes of the disaster. My ongoing research into this question eventually formed the core of *The Truth About Chernobyl*. In actual fact, on that June day I had already begun writing the story; I was simultaneously keeping a diary of events and tracking down elusive data.

Vatslav Mikhalski attended my lecture at Sovietski Pisatel. He listened silently for the full two hours, occasionally removing his glasses in order to wipe them dry. After the lecture he walked up to me, introduced himself, and said, "Let me have a manuscript of your prose. You do have one, don't you? Give me fifteen to twenty typewritten pages."

"Is this going to spend the next two years somewhere in a pile of manuscripts?" I asked.

"Of course not. Within a month we shall read the work

and make a decision. This subject matter gets top priority."

This hard-nosed Slav, the talented and stubborn Vatslav Mikhalski, kept his word. *A Moment of Life*, a book of stories relating to nuclear matters, appeared in 1988. Its publication was still some time off, however. Meanwhile . . .

**
*

I got another excited call from Aleksandr Isayev, the first deputy chief editor of *Druzhba*. He summoned me over to examine some interesting material that had come to his notice. The photocopies of newspaper clippings that I was shown when I arrived at his office were indeed interesting. I was invited to comment.

There was a photocopy of a cross-section of an RBMK reactor, with circles indicating the fuel channels. I was told that this drawing had been published in the Novosibirsk youth newspaper *Sem Dnyei* ten days before Chernobyl, and that if a Jewish calendar were placed over the newspaper page, reading from right to left, at the point corresponding to 25 April in the Gregorian calendar there was a symbol of explosion, in the form of an irregular six-pointed star of David. The two upper points of the star resembled the hands of a clock reading approximately one thirty in the morning. If the drawing was held up to the light, then on the other side, against the symbol of the explosion, the Cyrillic abbreviation *YCCP*, which stands for the Ukrainian Soviet Socialist Republic, became visible. In other words, the date, time, and place of the explosion were indicated. The newspaper text also referred to the issues of *Pravda* for 1 and 2 May, in which two carnations appeared next to the headlines of certain articles.

Photocopies of the headlines with the carnations were included. A pair of carnations is typically placed on graves. Then came the question: who stood to gain? Who knew about the explosion in advance? And who was being warned by the Novosibirsk youth newspaper? Then came the obvious conclusion: the only ones who stood to gain were the Yid-Masons, who warned their fellow Jews in Kiev to leave town ten days before the accident.

"What do you think?" asked Rakov and Isayev.

"Where did this come from?" I asked.

"From the KGB."

"Well, let them investigate it. It should be easy to follow the leads. Where did the drawing of the RBMK reactor come from? Who submitted it for publication? Was more than one emblem published by the newspaper? If so, how long ago did these emblems start being published? And so on. There's a branch of the Soviet Academy of Sciences in Novosibirsk. It may be the symbol of an institute, perhaps of nuclear physics. The KGB should check out any suspicions they have."

"What is your personal opinion?"

"It's nonsense. Yet if you consider this material from the anti-Semitic point of view, it must be advantageous for the Yid-Masons to harm the Slavic genetic pool. But such a narrow approach limits the impact of the Chernobyl disaster excessively. The consequences of Chernobyl go far wider and deeper than that. Radioactive harm has been done to the genetic pool of all mankind. If you consider the question of who stands to gain in global terms, there's really only one answer: Chernobyl has harmed everybody. The question can also be put differently: Was the Chernobyl explosion the result of sabotage? There are two answers to that. First, we could assume that a psychopath with specialized knowledge of nuclear physics and a strong grudge against mankind was involved. A kind of nuclear Herostratus of the twentieth century. He could be Russian, Jewish, Ukrainian, or a member of any other ethnic group.

"The other possible assumption is that the Chernobyl explosion was an act of sabotage by forces opposed to *perestroika*. Such people are usually poorly educated and incapable of foreseeing the results of their actions. Seven decades of Soviet history prove my point. These people tend to care mainly about hanging onto power and eliminating their opponents. Everything else can go to hell. Of course, these people fail to realize that in the case of Chernobyl things could get really out of hand. Such individuals could again belong to any number of ethnic groups.

"However, if we are talking about sabotage, what about the global sabotage committed by the totalitarian system itself, which acts as a colossal destructive force? That's where the roots of this tragedy really lie.

"As I said before, now that this version of events has surfaced, let the KGB look into it. Enough harm has been done already."

***
***

I might have found this episode too insignificant to mention, had it not come up again, unexpectedly, four years later.

Meanwhile, work on *The Truth About Chernobyl* was progressing. I had already completed some rough drafts. I was absolutely determined to find and question some eyewitnesses of the events on the night of the explosion. I needed to broaden the list of victims and participants that

I had drawn up on my trip to Chernobyl and Pripyat, and during my stay at Clinic No. 6 in Moscow.

Then one day Petrov, head of the equipment section at Chernobyl, appeared unannounced and of his own accord in my office at Minenergo, on Kitaisky Proyezd, 7. I recorded Petrov's story, but it provided no leads to other participants. I needed to reconstruct the night of the explosion from pictures and individuals—otherwise the story would not hold together.

Someone reported to the first (secret) section of Minenergo that I was collecting material on Chernobyl, keeping a diary, questioning witnesses, and recording what they said.

Colonel Vladimir Klimovsky, the resident KGB officer on the staff of the ministry (his post camouflaged under the title "senior expert"), made a comment to my immediate superior, Yevgeny Reshetnikov, who then called me in and warned me to be more careful.

"Naturally, you have to write about all this. But stay clear of KGB Colonel Klimovsky. He's a powerful man. You could get yourself into all kinds of trouble."

Thereafter I collected my material at a faster pace. I instructed Tatiana Savushkina, who worked under me in the nuclear power station section, to get in touch with Dontekhenergo, a firm based in the town of Donetsk, and to try to reach Gennady Metlenko, who had been in charge of the electrical experiment on the night of the explosion. If I happened to be in the office when the call went through, Savushkina was to connect me with Metlenko. Savushkina was well placed to make the call; she was in charge of Donbasenergostroy and the Yuzhno-Ukrainskaya nuclear power station.

I had found out about Metlenko from Mikhail Alekseyev,

deputy head of Gosatomenergonadzor (the State Committee on Operational Safety in the Nuclear Power Industry), who had spoken to him shortly after the accident.

Metlenko was a very important first link in a chain of names and events. At the time, the last names of all those involved in the tragic events of 26 April in reactor unit No. 4 at Chernobyl were being kept secret. Each of the survivors had been made to sign a pledge to remain silent. Haste was therefore essential.

In Clinic No. 6, where I arrived on 4 May to question survivors, I was unable to talk to any of them. The operators were all in critical condition. I just got a glimpse of some of their faces, darkened by a nuclear tan. The doctors, nurses, and a few of the relatives of those who had died offered me some information. All of this was vital, heartrending material for my book.

I had to move quickly.

I combed through the press reports about Chernobyl, occasionally finding an important detail. The press also confirmed the identities—still being kept officially secret—of Kurguz, Akimov, Dyatlov, Genrikh, Gorbachenko, Vodolazhko, Lelechenko, and many others. I repeatedly checked and confirmed the names of the heroes of Chernobyl so as to avoid mistakes.

Eventually, Savushkina found Metlenko and tried to connect us, but I was in a meeting at the time. She did talk to him herself, however, and she recorded the conversation. Metlenko had already been moved to a new section, so as to sever any old connections he had. But he was caught unawares by the call from Moscow, and what he said was revealing. I included his story and the transcription of his conversation with Mikhail Alekseyev in *The Truth About Chernobyl*.

The official clampdown on information about Chernobyl was becoming increasingly severe. I was working like an intelligence agent inside a hostile camp. I noticed every word said about Chernobyl, each unexpected twist in the events and their disclosure. Pieces of the story began to fit together, and gaps got painstakingly filled.

I was, of course, greatly helped by my knowledge of nuclear power, the design of nuclear power stations, and the special features of nuclear technology; otherwise I would have been unable to piece together the mosaiclike portrait of the disaster.

At a meeting in the main hall of the Collegium of the Minenergo in July 1986, presided over by the vice-chairman of the Council of Ministers, Yuri Batalin, I sat next to Leonid Voronin, who later became first deputy minister for nuclear power. At the time he was deputy director of the VNIIAES (the All-Union Scientific Research Institute for Nuclear Power Stations) and a member of the commission whose job it was to analyze the documents, logs, printer diagrams, and Skala computer printouts from Chernobyl.

KGB Colonel Vladimir Klimovsky kept changing places among the audience, which did not entirely fill the hall. The colonel listened and strained to detect anything unusual. He had the round, brazen face of a bully, with chilling pale blue eyes.

Moving into the seat behind Voronin, he asked, "How are things?" He was unaware of my identity, though he must have caught glimpses of my face at the Chernobyl headquarters and at various Minenergo meetings. Luckily, he did not know that the man sitting next to Voronin was precisely *the* Medvedev whose unauthorized actions in collecting material for a book on Chernobyl had been the subject of a comment he had made to Reshetnikov, deputy head of Glavstroy. I thus became an involuntary witness to a curious conversation.

The focus of the meeting was some distance forward, near Batalin. They were discussing the digging of an underground wall to prevent the radioactive water on the plant site from reaching the Pripyat River, as well as the initial plans for enclosing reactor unit No. 4 (the Sarcophagus).

"We still haven't been given all the papers," Voronin said quietly, no doubt afraid of the colonel.

"Have you identified those personally responsible?" the KGB colonel asked Voronin unrelentingly.

"Yes, there are a few names, but we still need to do some checking." Voronin mentioned a name, which I was unable to remember.

"Go ahead and shoot him!" the bloodthirsty colonel blurted out. "Shoot him!"

Klimovsky then moved on to eavesdrop and offer valuable comments somewhere else, while Voronin, with an embarrassed smile, turned to me and said, "Colonel Klimovsky is a very hard man."

Ludmila Arkhipina, of Soyuzatomenergo, came to see me on official business at Glavstroy, bringing with her an insert from *Sovietskaya Kultura* that contained a passage from *Sarcophagus* by Vladimir Gubarev. I promptly read it and became even more convinced that nothing would come of our endeavors if anything but the truth was written. The Chernobyl catastrophe was being whitewashed and coated with a glossy veneer of lies, fabrications, and science fiction, whereas what was needed was the truth. And a lesson. It was essential to reconstruct everything that actually happened. People needed a painful mental jolt that would at long last make them think.

When I asked Arkhipina whether anyone who used to work at Chernobyl was now working for Soyuzatomenergo, she mentioned the widow of one of the operators. I asked her to find out the woman's precise identity.

Soon she called me with the name of Lyubov Akimova, widow of Aleksandr Akimov, former late-shift foreman of unit No. 4 at Chernobyl. I called Akimova to arrange a meeting. Although she had already signed a pledge of secrecy, she did not refuse.

During several conversations I had with her, I learned where I could find several surviving witnesses to the events of that tragic night, among them Smagin, Davletbaev, Genrikh, Palamarchuk, and Tormozin. It was a real triumph.

I later reconstructed and repeatedly checked the list of passengers on the first and second flights to Chernobyl on 26 April 1986. I talked with all of them and recorded their stories. They included Vladimir Shishkin, deputy director of Soyuzelektromontazh (the All-Union Department of Electrical Assembly); Gennady Shasharin, former deputy minister for nuclear power stations; Boris Prushinsky, chief engineer of Soyuzatomenergo (the All-Union Department for Nuclear Energy); Mikhail Tsvirko, head of Soyuzatomenergostroy (the All-Union Department for the Construction of Nuclear Power Stations), and many others. My meetings and conversations with Razim Davletbaev, Viktor Smagin, and Oleg Genrikh—heroes who miraculously survived that terrible night at Chernobyl and told me the truth about it—were pivotal. Their stories were disjointed; I had to word my questions with extreme care and hold repeated interviews in order to clarify details and help them remember exactly what happened. Each of these heroes of Chernobyl told me only what he himself saw and experienced. It was left to me to piece together the various elements, so as to present an accurate account of what happened, in time and space, at the very beginning of the Chernobyl era.

The important thing was to interview people as soon as possible, while the events were still fresh in their memory, before their grief, resentment, horror, and anger had begun to fade, and while there was still a desire to perpetuate the memory of the dead, pay tribute to the heroes, and blame those responsible.

I had to work fast, as Klimovsky and possibly others were on my heels. I rummaged through the whole of the civilian and military press, gathering crumbs of information about the prowess of the airmen and fire fighters. I supplemented my findings and the knowledge I had gained firsthand in Chernobyl by questioning dozens of people who had actually witnessed these men's achievements. I drained literally every drop of information I could from Ignatenko and Prushinsky, Kamenev and Shasharin, Tsvirko and Zayets. And, gradually, the mosaic of discrete facts became a riveting picture of a monstrous nuclear disaster and of heroic feats of valor.

Writing the story took massive amounts of mental energy. While working on *The Truth About Chernobyl*, I sometimes popped a validol or nitroglycerine tablet under my tongue, as I had begun to suffer chest pains and often could not sleep. Haste, it was clear, was essential.

**
*

Savushkina came to tell me that she had been summoned to the head office of Soyuzatomenergo. Someone had reported her conversation with Metlenko—possibly even Metlenko himself, in order to mitigate his offense. He, too, had signed a pledge of secrecy, and he had inadvertently blurted out some information over the phone. It was even possible that their conversation had been bugged. But the matter wasn't serious. Savushkina had little to explain: she had talked with Metlenko on instructions from Medvedev, and she had also met with Lyubov Akimova. Medvedev had left for the Rovenskaya nuclear power station and asked her to check certain details with Akimova. What else? That was really everything.

I had ordered Savushkina to go to the classified section* of Soyuzatomenergo and say exactly what had happened. And I truly meant it as an order. Nothing terrible had happened. But that was not how the KGB saw things.

**

Tatiana Savushkina, senior engineer in the nuclear power station section of Glavstroy in the Ministry of Energy, summarized the situation as follows:

Soon after my phone conversation with Gennady Metlenko of Dontekhenergo, I got a call from Vladimir Ivanov, the head of the first section of Soyuzatomenergo, asking me to come in for a talk. I did not like the sound of what he was proposing, so I ignored the first two or three calls. I told Grigori Medvedev, who instructed me to comply. After the fourth call, I went down to see Ivanov on the third floor. He was sitting alone in a big room with a few tables in it. I sat down across from Ivanov and invited him to proceed.

---

* The classified section in all ministries and government departments represented the KGB.

"Do you know a man by the name of Grigori Medvedev?"

"I certainly do. And you must know him, too; he worked with you in Soyuzatomenergo."

"That's beside the point, I'm asking you whether you know him."

"Yes, he works with us, as deputy chief of the main production unit for nuclear construction."

"What is your relationship with him?"

"There's nothing like that, if that's what you're hinting at. I work in the nuclear power station section that he is in charge of. All of us, all of his colleagues, have excellent relations with him. He did the recruiting for the nuclear power station section, and he gave all of us our jobs, including me."

"Why have you been passing him information about Chernobyl?"

"He instructed me to talk with Gennady Metlenko of Dontekhenergo. Donetsk and the Yuzhno-Ukrainskaya nuclear power station are the areas I'm responsible for. Medvedev could have made the call himself, but he was busy and asked me to do it."

"I'm asking you why you have been giving Medvedev information about Chernobyl. You work in the nuclear section, where you hear things and you know things. You know all that precisely because you work here. This is the central government apparat, and such information must not be divulged to anyone. You are engaged in important, almost secret work. Chernobyl is a state secret."

"Medvedev knows more than you about Chernobyl. He is a nuclear expert, he's spent years working in the operation, planning the construction of nuclear power stations. He was in Chernobyl after the explosion. What I told him is not directly related to my job. I was carrying out his orders with regard to Dontekhenergo. All of this was purely human information, about people. There cannot be any secret in that."

"On the contrary, you're wrong there. I know things about

Chernobyl that even your Medvedev doesn't know. There's a lot he doesn't know."

"Then you should tell Medvedev," I said. "He's working on a book about the Chernobyl disaster."

"I'm not allowed to," said Ivanov. "You are free, Comrade Savushkina. But you will be called in again for another talk. And Medvedev, too." Then, after a moment's silence, he added, "I have a proposal to put to you. Tell us about the progress Medvedev is making on his book, about the channels through which he gets his information and his meetings with eyewitnesses to the events."

"He doesn't report to me!" I replied indignantly. Ivanov smiled; he's got some nerve. I blew up. "You should tell Medvedev everything you know. It's important for everyone, not just for him—for you and for me. Medvedev wants to tell exactly what happened. We need the truth about Chernobyl."

Then it was Ivanov's turn to blow up. "The truth about Chernobyl is known to those who are supposed to know. And those who are not supposed to know can just go on living the way they were before."

"But Medvedev is supposed to know."

"We'll take care of him. Who does he think he is? So he needs the truth! Huh!" He then starting warning me again. "You're going to get yourself into lots of trouble. This is none of your business!"

"You've covered up plenty in the past, before Chernobyl blew up. Medvedev wants to tell the truth so that such things will never happen again."

Ivanov smiled maliciously while I was letting off steam, and kindly allowed me to finish what I was saying. And then he once again explained, "Special people have been given the job of finding out what happened at Chernobyl. But you, Comrade Savushkina, acting on Medvedev's orders, have been doing unauthorized things. Don't do it again."

The conversation went on for about an hour. Twice someone

came in and distracted Ivanov. As I was leaving, he said, "Make sure that you say nothing to anyone about our meeting or our conversation. Understand?"

"Yes, I understand. The only thing I don't understand is what's so secret about our conversation."

"There is nothing secret, but don't tell anyone. And we'll be having Medvedev in here, too," he said emphatically.

This is too much, I thought, closing the door behind me. I never met him again, though I imagine he passed my name along to his superiors in the KGB.

Then, a couple of weeks later, I got another call, this time from Shugalo, of the first section of Minenergo, asking me to go and see him. I did so promptly.

Shugalo had a big pot belly. He got straight to the point. "You doubtless know why you have been summoned here?"

"I can guess."

"I must ask you to write down everything you have told Medvedev," he said, almost as if issuing me an order.

"I've done nothing wrong, and I'm not going to write anything."

"In that case you're going to get into trouble. You could lose your job. If you've done nothing wrong, then all the more reason for you to write this information down, since it is nothing out of the ordinary. Then it will all be handled through the appropriate channels."

"You're making up a file on Medvedev?"

"Just write! Go on!"

I decided to write down what had happened. In any case, they had been bugging the phones. They must have listened to my conversation with Metlenko. And I had also spoken with Lyubov Akimova. She had been restrained; she had told me that all the people involved in Chernobyl had been summoned to the KGB and warned not to talk to anyone about what had happened. Everyone was required to sign a pledge of secrecy. It's possible that Akimova herself had reported our conversation to the first section. I don't know. For these reasons I decided to describe the substance of my conversations with Metlenko and Akimova.

My "informer's report" filled two sheets of paper. It was all very unpleasant and shameful. I felt terrible about making life difficult for Medvedev, really ashamed. I was certain that I was in the right. Nonetheless, I wrote my report: whom I had telephoned, what I had said, why, and for whom. I argued in my report that it was necessary to do all these things. I just wanted to finish writing the report and hand it in and never meet these people again.

Shugalo said that he was a lawyer and that I could be taken to court. I was beginning to wonder about that myself. It was terrible to think that I could be fired; I would have a hard time getting another job.

After I had written my "informer's report" I felt really bad, as if I had stooped to something terribly underhanded and would never be able to extricate myself. What also occurred to me was the possibility that nothing would come of it, that they could prove nothing, that they would not concern themselves over such a minor matter. But they *would* have me fired. That they would do.

A few days later the head of the directorate called me into his office. I immediately knew something was up when I looked at his sour, stony face. He laid into me right away. "Why have you been doing it? Medvedev's book won't ever be published, and you're going to ruin your life."

I interrupted him. "I've already been in the first section and signed an 'informer's report.'"

He stopped abruptly. Then he said, "I hope that's the first and the last time."

"I hope so, too." And then I left. For some time after that I was constantly expecting to be summoned or to get a phone call. . . .

When Savushkina told me about these conversations, I tried to reassure her, saying that they had no grounds for victimizing her. I advised her to stay calm and confident. "The people who were talking to you are cruel and not very bright. God is their judge," I said as I took leave of her, after thanking her for being so helpful and apologizing for any alarm I had caused her. "If we manage to publish this book, you'll be the first to get a copy."

I met with Yevgeny Ignatenko, who had just returned from Chernobyl. His puffy face was pale and powdery. He had put on a lot of weight: radiation is very conducive to the retention of fluids. I felt sorry for him. Boris Shcherbina, deputy chairman of the Council of Ministers, was holding Ignatenko formally responsible for Chernobyl. The fact is that at the time of the explosion Ignatenko was deputy head of research and development at Soyuzatomenergo and was responsible for the equipment used in nuclear power stations. Therefore he was, in a sense, at least indirectly to blame for the Chernobyl accident. He had been negligent; he had used an unreliable reactor in a nuclear power station project.

Ignatenko had received a dose of about a hundred ber. But how much radioactive contamination had he breathed in? He had now come to Moscow for treatment at Clinic No. 6.

"How are things, Yevgeny?" I asked as gently as possible.

"Things are only just beginning. The main part is yet to come."

"You here for long?"

"Shcherbina won't let me stay long. I'll just get some treatment."

He sounded resentful. Shcherbina had mercilessly nailed him, like a butterfly on a pin.

But victims who had received 25 ber had to be evacuated from the zone.

As I watched his portly figure recede along the corridor, it occurred to me that all the efforts he and I had made to warn people about the danger had failed to elicit a timely response from either the journals and publishing houses or the government organs that foolishly delayed the publication of "The Expert Opinion," "The Reactor Unit," "The Hot Chamber," and "A Nuclear Tan" and impeded anything that might have alerted people to the impending catastrophe.

When I had met Ignatenko at Chernobyl, he had said that if "The Expert Opinion" had been published before the explosion it would have been a best-seller. He was, of course, exaggerating. I wondered whether the explosion might never have happened, still naïvely believing in the magic power of the creative word, of the truth. But was that so naïve? After all, the Word was God. And if God is the highest justice, the Word would have prevented and stopped the calamity. The Word can materialize. I believe in the immense power of words; that's why I fight.

"The Expert Opinion" was published in November 1986. As I saw for myself, it landed at Minenergo like a bombshell. I remember a meeting of the enlarged collegium of the ministry at which many of those present were reading "The Expert Opinion," recognizing the prototypes and exchanging opinions. They all agreed on one thing: if "The Expert Opinion" had been published in February 1986, there might have been no explosion at Chernobyl.

But what was fated to happen did happen.

The trouble with people is that they are unwilling to believe prophecies, even those that serve as warnings to them, and this lack of belief causes them to compound their errors. Everyone goes his own way. Each people has its own dignity . . . or should we say that each people wants a happy destiny, but has to endure great suffering, often being left with bleak prospects.

It is so important to raise the culture of a people, and the culture of its rulers or its fully empowered representative of authority.

The administrative-command system is primarily guilty of stultifying, blinding, and spiritually robbing the people, stripping away its sense of danger and self-preservation. The one hundred and ten million murders committed by Stalin certainly had their effect. Chernobyl is a continuation of genocide, only at a qualitatively new, more devastating level. May God help the crippled peoples to survive, and, over the centuries, to make good the harm they have suffered; may He also instill in them the power of reason and the will to achieve a healthy and sensitive life.

Meanwhile, the cover-up of Chernobyl was of mounting concern. We had to hurry. My work on the book was progressing slowly. More and more clarifications were required as the story unfolded; I repeatedly returned to eyewitnesses, to those involved in the events of that terrible night.

"I respect your desire to write the truth," Smagin told me. "But mark my word—the story is not going to be published."

"Shcherbina would call us in," Davletbaev confessed. "Again and again he got us to promise that we would say nothing to anyone. He showed us the draft of the order from the Council of Ministers banning the disclosure of information about Chernobyl, emphasizing that such information must pass through the Gosatom commission on Chernobyl. If they really get rough with us, we shall be obliged to disown our testimony."

That's fine! I exclaimed inwardly. Disown it: it won't change a thing. The story is almost ready. And I shall accept full responsibility for it, so relax!

It was too late to stop the story now; it already had a life of its own, awaiting its readers. And it would reach those readers.

*

We had to make sure that "The Reactor Unit," an extremely topical story that had been ready for some time, was actually published. Late in November 1986 I sent it to *Neva*, only to be told that they again wanted the stamps of the accursed nuclear censors. It was as if Chernobyl and *perestroika* had never happened. The government departments were still strong, and they had *glasnost* by the throat. What could I do? Once again I went as a supplicant to Prushinsky at Soyuzatomenergo. After he had stamped the title page I sent it to Boris Nikolsky at *Neva*.

Prushinsky told me that Shcherbina had issued an order whereby all Chernobyl material was to be channeled through the Chernobyl commission. He concluded, "Your story is now at a dead end."

"It doesn't matter. We'll make it," I replied.

In February 1987 "A Nuclear Tan" was published in *Ural*. Of course, it could and should have been published earlier, before Chernobyl. The strength of the story's impact became evident from the excited reaction of the staff, particularly the operators, of the Beloyarsk nuclear power station; the station's management, on the other hand, heartily deplored the story. There was press coverage followed by letters from readers. It would certainly have helped if the story had been published before the event, or even just after. Now it seemed too late. But was it? Is a calamity of which nobody is aware not yet a calamity? Even then people need to be made aware of its full magnitude.

Lies were everywhere: Goskomgidromet (the State Committee on Hydrology and Meteorology), Minatomenergo (the Ministry of Atomic Energy), Minsredmash, Shcherbina, and Marin (the Central Committee's expert on nuclear power)—all of them lied. The creators of Chernobyl were laying down a thick web of lies upon the radioactive earth, assuring people that it was clean, healthy, and habitable. "Go ahead and live there!" they were saying. "It's even more advantageous than without radiation."

Pseudoscientists were issuing pseudoscientific reports proving that resettlement was out of the question, the government could not afford it. These scoundrels just loved to hide behind the state that they themselves had made, thus safeguarding their own skins and the whole of their mediocre burnt-out lives. Now, however, it was too late for them.

**
*

I went to see Vitaly Samchenko,* deputy director of the scientific and technical department of the ministry, who was now in charge of translating from various languages the technical documentation obtained through industrial espionage. At one time we had worked together on

---

* The name has been changed.

Minsredmash reactors; something must have happened in the intervening years. Samchenko was a good physicist with a knowledge of several foreign languages. But now he had switched from nuclear technology to this new occupation.

"Is the pay any better here?" I asked him.

"A little bit," he replied evasively.

"And how's the work? Are any blueprints coming in?"

"They certainly are! How could they not come in?"

"Where do they get them from?"

"We have people over there with the capitalists; they steal them from the design studios and factories."

"Do we buy them for hard currency?"

"Hardly—nobody wants our roubles."

"Does it do any good at least?"

"Some, but not much. Over there they have different production standards and different technical requirements in industrial design. You know that yourself. So we have to reprocess the stuff, switch it from European to our own Russian rails, that sort of thing. As you might expect, much, if not all of it is lost. So the net gain to us from this stolen material is not more than 20 percent. By the same unofficial channels, for example, we managed to get hold of a heat sensor for Chernobyl"—he showed me the instrument—"to enable us to measure the temperature in the destroyed reactor No. 4 from two or three hundred meters—from a helicopter, for example."

"Couldn't it have been bought through official channels?" I asked.

"Yes, but we have gotten so used to getting things unofficially—it's faster," he said with a laugh.

So that's how we had been working. We could not even apply stolen designs and ideas. . . . No! *Perestroika* was as necessary as the air we breathe! Gorbachev had started something really worthwhile!

*

It seems that after the publication of "The Expert Opinion" in the Soviet-Bulgarian journal *Druzhba*, some probing inquiries began to be addressed to its editors, who were increasingly asked all sorts of questions about me, what kind of a writer I was, whether I was well known, and so on. These questions could, of course, have had some other purpose. "The Expert Opinion" touched on the interests of two powerful government departments: Minsredmash and Minenergo, which together comprised the military-industrial complex. The author needed to be taught a lesson. But how? It was now 1987, the second year of *perestroika*.

I once had a call at work from Rakov.

"Grigori, I was wondering whether you might address our Politklub. Members of the Komsomol meet after work and ask writers all sorts of loaded questions. But I'm sure you'll handle it just fine. How about it?"

"Will there be any KGB agents present?"

"Where are they not present? They're all over the place, my friend. Ha ha!"

"Well, if I have to, that's all right."

"Good," said Rakov with a sigh of relief. "Take down this phone number. The name of the organizer is Leontiev."

I phoned Leontiev.

"Yes, we've been expecting your call. We shall send a car around for you anywhere in Moscow or the Moscow region. It'll be a Volga. You'll be taken where you have to go. We'll sit down and chat."

"There's no need to send a car," I said, "I'm not such a bigshot that I need to be driven around in a Volga."

"But it's no trouble at all!" said Leontiev excitedly.

"Just give me the address, the name of the subway station, and some general directions. I'll come."

"You know, we don't give out the address of our organization. It's really hard to find this place, there are all kinds of courtyards you have to find your way around."

You fool, I thought, that's really crude. You should have thought up a better story than that.

"I shall find it," I said. "Let me have the address; otherwise forget it."

"Let me consult on that a minute," Leontiev said, his voice trailing off. He was away from the phone for quite a long time, during which I could hear distant voices, a drunken shout, some wheezing, and a hissing sound in the receiver. When he came back, Leontiev's voice was gentle, insinuating. "You know, they wouldn't allow me to give out the address."

"Have a nice day, Ivan Petrov!" I said and hung up.

I then called Rakov.

"Listen, Aleksei. What kind of a Politklub do you have, huh? They don't give out their address. It's a mysterious organization. That couldn't be the place where authors get their balls stomped on, could it?"

"Ha ha! Come on now!" Rakov sounded embarrassed.

"So the Komsomol volunteers, the dirty rotten stool pigeons won't give out their address, huh?"

"No, they won't."

"Well, then to hell with them. That's their business. If they don't want the writer, too bad for them."

What kind of person are you, Aleksei Rakov? I wondered as I put down the receiver.

135

**
*

When I received the author's copies of *The Operators*, which had just been published by Sovremennik, I gave the first ones to Davletbaev and Smagin, from Chernobyl.

Both men thought very highly of the book. With a smile Davletbaev said, "Now I know you much better. We just had a nodding acquaintance in the past, although I did give you my testimony. Now I believe you, and the book helped. It's a good, warm book, the kind of thing nuclear power people would understand."

Davletbaev was right, and I was grateful to him. A book that tells the truth brings people closer together. He certainly trusted me much more as a result of it. I gave him the night scene from *The Truth About Chernobyl*, which I had just completed. He, of course, was one of the heroes of that night at Chernobyl. After reading the scene carefully he made a number of comments. "I don't believe it will be possible to get a story about Chernobyl published. There are some very powerful obstacles. Shcherbina recently got all the survivors together and warned them not to tell Medvedev anything about what had happened. He had us promise orally to remain silent. And we had already signed a pledge of secrecy earlier." Davletbaev laughed. "But it's a bit late now. I understand the book has been written. We all very much want it to be published. Let's hope it works out. And there's another thing: Shcherbina issued a totally secret order from the Council of Ministers that he himself approved. It says that no new work, report, or article on Chernobyl may be published unless it is first submitted to Shcherbina's Chernobyl commission. Actually, Verkhovykh, deputy minister of Minsredmash, is the real chairman of the commission. Anything on Chernobyl that comes their way they chop to pieces."

"We'll get around all that."

"But how?" he asked. "There's just no way it can be done."

"Yes there is: Gorbachev."

Bad news from Nikolsky at *Neva*, in Leningrad: the censors were not satisfied with the Minenergo stamp; they also wanted a stamp from Gosatom. I remembered Ignatenko's warning that the Soyuzatomenergo stamp would not be enough. The zealous custodians at Glavlit trust only Gosatom. Of course, its stamp had been in existence for thirty-five years, twice as long as the stamp from Soyuzatomenergo. The Leningrad Glavlit people, as true Leninists, were not to be fooled so easily.

What a country this is, with all its official stamps everywhere! It is truly the "land of slaves, land of lords" mentioned by Pushkin—a land of bouncers and bureaucrats with one thing in mind: prohibition. I naturally realized that stamps of approval were being placed not only on me and my works by the myriad censors of the vigorous empire established by Stalin, the Father of the Peoples. Not a single work could be published anywhere in the USSR without the censors' stamp of approval.

I began to conduct a mental review of everyone I knew who had worked with me at various times on the nuclear reactors of Minsredmash. I could think of many acquaintances, but no one I was really close to. On the other hand, there was Ivan: if he was still there, he would probably be willing to help me. We had held similar views since the sixties. But even if he were to affix the Gosatom stamp, it would still be unofficial, bypassing the established procedures. He would mix up the registration or affix the wrong signature. Glavlit had models of all signatures. There was also the possibility of trying the recently established Minatomenergo, although it was unlikely that Glavlit would take it very seriously, regarding it as a newcomer to the field. So . . .

The Truth About Chernobyl was ready, in typewritten form. I was wondering which journal I should take it to. Throughout the USSR a cordon had been thrown around the truth about Chernobyl. Shcherbina had erected some daunting obstacles. Extracts from the secret banning order issued by the Council of Ministers had been sent to the editors of every journal and publishing house. I could visualize à sly smile on the smug, pock-marked face of the deputy chairman of the Soviet government, Boris Shcherbina.

I could also imagine him making a defiant gesture at me and saying, "You'll get nothing out of me! *No pasarán! Patria o Muerte!*"

I could also hear Rakov's memorable words—"The Soviet authorities are smart! You're smart, but they're even smarter!"—although he did print "The Expert Opinion." Now, however, in 1987, he would not have published it. But the deed was done, and I was very grateful to him, whoever he really was. The Word works in inscrutable ways. And in "The Expert Opinion" the Word was certainly doing its job.

"The stamp can only be unofficial," I said in response to Nikolsky. "Neither of us is going to get a stamp from Gosatom by sending them 'The Reactor Unit.'"

"All we need to do is to show the stamped story to our

own dumb censors, and I think it will be OK. If it's not good enough for them, we shall take it right back, and then together we can figure out what to do next."

I found this acceptable. When I called Ivan at work, I was told that he was away on leave. It had been a very long while since we had met; we had spoken a couple of times over the phone in the previous two years. He had called me once, after reading "The Operators" in *Literaturnaya Uchoba* in 1981. And I seem to remember we once attended the same meeting at Minsredmash, but that was it. I wondered whether he would agree to help. After all, it was now 1987, long after Chernobyl. Although the nuclear power people had been subdued for a while, they were clearly beginning to bare their fangs, as could be seen from the wave of prohibitions.

"Hi, Ivan! You haven't forgotten Grigori Medvedev yet, have you?"

"Oh, glad to hear from you, Grigori! It's been a long time. I bought your little blue book at the Atomizdat shop. It's called *The Operators*, published by Sovremennik. I liked it a lot; it reminded me of when we worked together on the reactors. But don't expect everybody to like it. It's time we put an end to our nuclear secrets. They're a damned nuisance. We were getting what we deserved even before Chernobyl."

"Well, actually, that's what I wanted to talk to you about. Nikolsky at *Neva* wants to publish my story 'The Reactor Unit.'"

"I read your short stories in *Neva*. They've now been put in the little blue book. Gosatom has been keeping a close watch on your publications. They're jealous and spiteful. Odoyevsky and Yeremeyev are tearing their hair out. What a thought! Medvedev managed to go ahead without their stamps. Appalling! You know they've plastered their stamps all over Russia. We've cut ourselves off from the people with this damned stigma. There was never censorship like this in tsarist Russia."

"Well, that's just it, Ivan! That's what I had in mind. There's no limit to the harm your Gosatom can cause."

"At one time it was also your Gosatom. Don't you remember, reports on your reactor were sent to Department 16 at Gosatom?"

"Yes, I remember, Ivan. Nikolsky at *Neva* is asking for the stupid Gosatom stamp. I told him that I had a friend at Gosatom but didn't mention the name. In this thing, the others are our sworn enemies. The big question is: can you put the stamp on 'The Reactor Unit'?"

Ivan said nothing for quite a long time. I was so embarrassed that I began to blush.

"But I'm with Odoyevsky's press section," Ivan replied slowly. Then abruptly his voice brightened. "Wait a minute, there's Smirnov from Melekess, remember him? Maybe he's the man. He's empowered to do this. And we can leave Dim Dimych alone: he'll hang himself."

"Leave Smirnov out of this. He's a special case. I remember at Melekess he was red in the face, strutting around all the time. You couldn't have any meaningful contact with him at all. But why don't you see what you can do? We've got to knock a hole in the armor around nuclear secrets."

"OK, I'll try, but keep it quiet. If anything goes wrong, I'll sign the stamp myself. Zhenya Kulov has usurped the right to place stamps on Chernobyl manuscripts. Or, rather, to block their progress. Right now he's extremely resentful: he was kicked out of the party and lost his job as chairman of Gosatomenergonadzor. He might well go crazy. He's an extremely ambitious man. How do you like that? You see, Chernobyl wrecked all their careers, so they're going to do whatever they can to make it seem as if Chernobyl never happened. That's what they're doing now."

Ivan eventually applied the stamp. I'm not sure who signed it. As I understood Nikolsky, all that mattered anyway was for him to be able to wave the stamp in front of our dumb Leningrad censors.

The stamp had been applied somewhat unlawfully, but the question arises as to whether its very existence was lawful. Or was the administrative-command system, in defense of the power it had usurped, shielding itself from the poor disenfranchised people and imposing on them an inherently hostile ideology that had entangled the country hand and foot in a sticky web? Was the administrative-command system itself moral and lawful?

It would seem that we had done it. But I was still in a somber mood. I went to Minatomenergo to see Ignatenko, who had just returned from Chernobyl. He told me of a meeting he had attended in Verkhovykh's office at Gosatom at which they had discussed me. A decision had been made to keep track mainly of my writings as a professional nuclear expert. Publication of the rest was to be allowed, as the stories were nonsense anyway and did not trouble Gosatom. The appearance of *glasnost* could thereby be preserved.

"They're paying too much attention to me."

"They'd like to get their hands on your Chernobyl manuscript."

"Let them."

When I went to see my former boss, Yevgeny Reshetnikov, who was then deputy minister for nuclear power, I offered him a copy of "The Reactor Unit." I was grateful when he agreed to read it. He quickly approved it. I asked him to place the Minatomenergo stamp on the manuscript and sign it. In this way I would provide extra backing for Ivan's Gosatom stamp.

"No problem," said Reshetnikov.

Ten minutes later the story's title page was embellished with the stamps, the signature, and the date.

"Have you finished work on Chernobyl?" Reshetnikov inquired.

"Almost. A few details have to be cleared up. But I'll do that as I go along."

"You know, Shcherbina is really worried about your book. He's afraid you might say something in it that might suggest he was against evacuation."

"But he was. And many people know it."

"That's the whole point," Reshetnikov emphasized. "Boris Shcherbina told me that he is also writing a diary of Chernobyl and making notes about that period. If you agreed to give him your manuscript, he would share with you his material; and, of course, he would give you the green light for your own work."

"He's a dangerous sort of sponsor," I said, laughing.

"But he'd be a good one, too." It was now Reshetnikov's turn to laugh. "You think about it. A proposal of the deputy chairman of the USSR government counts for something."

"Let me think about it," I said as I prepared to leave, so as not to offend him.

*

The question of where to take *The Truth About Chernobyl* was more problematical. After much thought I decided the book should go to *Novy Mir*, where there were two men to whom I could entrust the manuscript, having first looked them squarely in the eye. They were Anatoly Strelyany, who had been the first editor of my book at Sovietski Pisatel and was now in charge of current affairs at *Novy Mir*; and Sergei Zalygin, whose fearless and stubborn defense of the purity of his native soil and whose valiant fight against the molelike depredations of Minvodkhoz were worthy of the greatest respect.

I went to see Anatoly Strelyany in July 1987 and laid the manuscript of *The Truth About Chernobyl* on the table before him. He eagerly undid the tape around the folder and started reading a few passages at random.

"This is atrocious! How could they be so lousy?" Strelyany exclaimed. Then suddenly, looking up at me with his pale eyes, he added, "Have you got the nuclear censors' stamp?"

Et tu, Brute?! I exclaimed inwardly. Does this mean that even the fearless Strelyany wants the devil's stigma?

"Let me just go talk to Vidrashka"—first deputy chief editor, in charge of censorship—"and get to the bottom of this."

Upon returning a few moments later, he said, "I knew it. You've got to have the Gosatom stamp."

"To hell with the lot of you," I said angrily.

I took back my manuscript and went to the Sovremennik publishing house to see Leonid Frolov, who seemed to respond more favorably to my entreaties. On the spot he appointed Lyubov Kuleshova as editor of the book. She was a fearless woman. At first we squabbled when I tried to lay down certain conditions: apart from her, nobody was to read the manuscript; it was to be kept securely in one place; no copies were to be made; and so on. By the time she had finished reading it, however, her views and mine coincided. I told her that soon Sovremennik would be receiving, if it had not already received, an extract from Shcherbina's order demanding that all material related to Chernobyl be submitted to Gosatom. This prospect did not seem to worry her unduly. She promptly handed *The Truth About Chernobyl* over to Valery Zalivaka, the section chief. He read it and fell sick, as did his wife, from the experience. More importantly, the book was approved for publication. It now had to be printed as soon as possible. Typesetting and layout were completed at lightning speed, and the manuscript received all the signatures it needed for publication.

Then came the extract from Shcherbina's order. To Lyubov Kuleshova's great regret I took back the manuscript of *The Truth About Chernobyl*, because it looked as though a raid might be imminent.

I phoned Nikolsky in Leningrad to ask whether he would be interested in reading *The Truth About Chernobyl*. When he said he would, I hopped on a train and took the manuscript to Leningrad. But Nikolsky then began to hedge, saying that Yuri Shcherbak had already completed a book about Chernobyl and that he, Nikolsky, had to read that first.

"Shcherbak did his job and I've done mine. Apart from anything else, they were clearly not out to get *him*. Come on, Nikolsky, who needs this kind of legalistic nonsense!? Good-bye!"

At this point I remembered the courageous Slav Vatslav Mikhalski. I called him at Sovietski Pisatel.

I took him the text and then went for a week's rest at

the seaside. When I got back, I found that not only had the manuscript been read by the entire editorial staff of Sovietski Pisatel, but it had also been approved and sent to the printers.

"What do you think of it?" I asked Mikhalski.

"I can't put it into words," he replied.

The text of "The Reactor Unit" had been composed at *Neva*. The first issue for the following year, which was to include my story, was awaiting a signature so that it could be printed. Suddenly there was a telegram from Nikolsky. I called him, only to be told that the censors were stopping "The Reactor Unit" because the Gosatom stamp bore the wrong signature.

Ivan must have signed it himself, I thought. And the censors are fools, all right, but they still managed to sniff this one out.

"So what?" I asked. "You yourself, Boris, said that all you had to do was wave the stamp in front of the stupid censors' noses. But it turns out they're not so stupid. I mean they are stupid, but . . . Why did you give them the manuscript with the stamp? Couldn't you have just waved the stamp in their faces and then taken it back?"

"Vladimir Solodin, deputy head of Glavlit, came here himself. A commission of investigation has been set up. You're going to be prosecuted." In his agitated state, the burr in Nikolsky's voice was becoming more pronounced than usual.

"Go ahead and prosecute me, you jerks! I'm withdrawing the story. Let me have the manuscript back immediately!"

"But it's been composed already. You've really let us down."

"So you've composed it—now decompose it."

"Who signed the stamp?"

"A friend of mine."

"What's his name?"

"I'm not telling. Blame it all on me! And give me back my manuscript."

"This is terrible. This is all wrong. Why don't we both go to Gosatom and persuade them?"

"They hate me down there. And they hate you, too. They're just in it for their own careers. You, on the other hand, produce literature. I mean you're not publishing a stamp, you're publishing a work of literature."

"But we can't do it without the censors!" he wailed. "Look, are we or are we not going to Gosatom?"

"We're not! Forget it! Let me have my story back, that's all!"

"The commission has been set up, and they will be questioning all journals and publishing houses that have your manuscripts. They'll make sure that you never get into print again!"

"Come on!" I replied. "They'll never make it. It's too late now. I've already got into print, and I'll get into print again."

In January 1988 Sovietski Pisatel made up the proofs of *The Truth About Chernobyl*. Just after they had been read and checked for errors, the editor announced that a letter had arrived from Gosatom. Pursuant to Shcherbina's order all Chernobyl materials were to be sent for approval to Verkhovykh's commission.

I immediately composed a statement addressed to Igor Skachkov, chief editor of the Sovietski Pisatel publishing house. "I hereby withdraw the manuscript of *The Truth About Chernobyl*. Please deduct any expenses incurred by the printers for typesetting the text of the manuscript from the fees due to me."

There was, naturally, an uproar. But the cordon around me was tightening, and I had to focus on breaking out of my encirclement.

*

By now I was really depressed, and I stayed that way for another couple of weeks. They were closing in on me. What should I do? Where was the way out? I could not abandon what I had already started. But who would support me? On the night of 17 February 1988 I lay down to sleep, utterly exhausted. That night I had a most unusual dream: I was lying on my back on the ground and looking up at the pitch black starry sky. The stars were twinkling, sharp and clear. Suddenly they began to rush from all over the sky toward its highest point, where they formed two beautiful starlit faces, a man and a woman, with huge, bottomless light eyes. The man's face started shrinking and descending, until it came to a halt some thirty meters above me. It stared straight at me. Then, abruptly, a slightly muffled, cavernous bass voice said, "And how is life on Earth?"

"So-so, nothing special. We are fighting," I responded, showing no surprise.

"Wouldn't you like to come up here with us?" (I remember clearly that he said not "into the sky" but "up here.")

"It's a bit early for me," I answered, as if it were not God's business to decide whether it was early or late or just right. "I still have work to do here on Earth. It's the only one we have in the universe."

"There are plenty of planets like Earth in the universe," the voice said. "So you won't come and join us?"

"It'a bit early for me," I answered the Lord confidently.

For a while the starry male face remained above me, its enormous clear eyes staring searchingly at me. Then the face rose in the sky, growing as it did so, until it stopped next to the woman's face.

The woman's face began to shrink and descend to the same altitude, where it stopped above me, staring at me probingly and lovingly, without a word. Compassion was visible in its eyes. Soon the face soared back into the sky, and after a moment both faces vanished among the stars.

When I awoke next morning I was happy and relaxed, my heart filled with joyous energy. For three or four weeks after that, wherever I went I was accompanied by two beautiful faces, my guardians, at my right side, roughly level with my heart.

In May 1988 my story "A Living Soul" was published in *Ural*. In July my book of stories *A Moment of Life* was put out by the publishing house Sovietski Pisatel, without *The Truth About Chernobyl*. That same month I received a summons from the Moscow military prosecutor on Rogova, 6 (Military Unit No. 3363), for purposes of investigation. When I phoned him, the prosecutor, Aleksei Khalyavchenko, asked me to come in without delay.

I sat and waited in the corridor of the prosecutor's office. A soldier was seated on either side of the entrance, each with a Kalashnikov across his knees. A tall, solidly built soldier with a stony mien paced to and fro along the corridor. The barrels of the automatic weapons were trained on him.

"And for that I have to waste my young life? They're going to put me away for years," he said to himself nervously as he paced up and down, up and down.

"What have you done?"

Then they took him away.

Khalyavchenko arrived. He was a well-built Ukrainian with a large face and cunning, round eyes. He stared at me inquisitively.

"Come in."

As soon as I entered his office I tried to win him over to my side, telling him about the vital importance of the work I had done to familiarize readers with nuclear matters. I pointed out that if the public was not made aware of these things we would all perish, and in short order—prosecutors and all.

There was a typewriter on the table in front of him. So far he had not used it; he was listening with apparent interest, and with the glint of a smile in his round, dark eyes.

After a while I felt little desire to continue, even though the need to do so was obvious. Despite all my efforts to enlighten this ignorant prosecutor, he just sat there silently, as if he had not understood me. There seemed no point in wasting my energy on him.

More silence. Then he said, "Some material has come in about you." He removed a copy of "The Reactor Unit" with Ivan's stamp on it from a desk drawer. "Tell me, who put this stamp on here?"

"Someone who works for Gosatom."

Khalyavchenko inserted a sheet of paper into the typewriter carriage and started typing.

"His last name?"

"He's a friend of mine. Unfortunately, I cannot give his name."

He typed some more.

"Where are the manuscripts of your works?"

"They are no longer where they used to be: they've been published."

"Where were they published?"

I listed the journals and publishing houses.

My prosecutor looked at me with curiosity, even with a certain conspiratorial air. He typed some more.

"Who put the stamps on those works?"

"According to regular procedures, Ignatenko, Kamenev, and Prushinsky."

They doubtless already knew all those names. He kept on typing.

"What are their official positions?"

I listed their positions in Soyuzatomenergo. They no longer worked there, and I did not know their new positions. All the stamps and signatures were legal. Everything was above board, and I was not getting anyone into trouble. Khalyavchenko now pulled from the drawer some of my manuscripts bearing the Soyuzatomenergo stamp.

"Are these the ones?"

"Yes they are," I replied as he continued typing.

Suddenly he stood up, walked over, and sat at a small table near me. Leaning forward and glancing sideways, he said quietly, "What kind of criminal are you? You saw that fellow outside in the corridor? He really is a criminal. Come back tomorrow with a statement from your house management people certifying that you have a clean record, and with that we'll close the file on you."

The phone rang. Khalyavchenko picked up the receiver.

"Yes, I've also read Gromyko's memoirs. They're really rather boring. Medvedev's story is much more interesting."

Khalyavchenko walked back to his desk and sharply turned the handle of the typewriter carriage, noisily ejecting the page on which he had typed his questions and my answers. He asked me to read it; I found it all in order and signed it. We said good-bye.

The next day I returned with the statement Khalyavchenko had requested. And my dealings with him ended there. I wonder whether he has read *The Truth About Chernobyl*. In any case, I wish him well and hope that he will no longer do the bidding of the censors and Gosatom.

I realized that if they had had any incriminating evidence they would have put me behind bars; and Comrade Khalyavchenko must have known that, too. But there was no evidence, despite the best efforts of Nikolsky, Glavlit, and Gosatom. They had collected all my proofs. It would be interesting to know who supplied the prosecutor with "The Reactor Unit"—Glavlit or the editors of *Neva*? It no longer mattered, however; all that mattered was that there was no incriminating evidence. Nor could there have been any.

On the other hand, a crime had been committed by Minsredmash, as the soil and people's lives had been wrecked by radiation. These crimes had been kept secret, jealously guarded by our censors. And by our prosecutors, too. Isn't that so, Aleksei Khalyavchenko? Isn't that so, Lilia Khokhlova?

**
*

On my way to the prosecutor's office I already knew that for several months before his summons Khalyavchenko had been visiting all my previous places of employment, searching for incriminating evidence and failing to find any. He had also been to the various publishing houses, where he had tried unsuccessfully to get his hands on the manuscript of *The Truth About Chernobyl*.

Leonid Frolov, director of the publishing house Sovremennik, had received calls from the Communist Party's Central Committee, the Soviet Council of Ministers, and Gosatom, all of them frantically hunting down *The Truth About Chernobyl*. They didn't find it, nor could they have. The publishers rose to the occasion and did not sell me out. There are more honest people, Comrade Khalyavchenko; far more than there are scoundrels, Comrade Shcherbina; and far more than ambitious incompetents, Comrade Mayorets.

PART THREE
Frontal Assault

ON 7 JULY 1988 I set out once again, though without much hope, for the editorial offices of *Novy Mir*. The situation seemed desperate, with censorship rampant. Even so, I had resolved to try Zalygin, with whom I had formed some barely perceptible yet solid bonds. When times are hard, people look over the edge of the abyss and then, casting off fear and degrading caution, do the one thing that is essential for the preservation of life.

Luckily, Zalygin was in his second-floor office when I arrived. I placed the blue folder containing *The Truth About Chernobyl* on the desk before him.

"Here is a manuscript about the Chernobyl disaster. Will you have a chance to read it?" I asked him.

"Yes, I'll read it. Only I can't promise that I will do it right away. It'll take me a couple of months to get to it."

I left the manuscript with him. A few days later I happened to phone Margarita Timofeyeva, head of *Novy Mir*'s prose section, on some other business. She surprised me with good news. "You know what? Zalygin got us all together yesterday evening and started to describe the contents of *The Truth About Chernobyl*. It seems he opened your manuscript yesterday at home and read it straight through, until four in the morning. He spent a couple of hours telling us all about it; then, with a wave of the hand, he told us that we ought to read it. He left your manuscript with us in the prose section with orders to prepare it for composition. 'It's going to be our best publication of the year,' Zalygin wrote on a piece of paper attached to the title page. I gave the manuscript to Inna Borisova for her to read. Do you have another copy for me, so as not to waste time waiting, so that both of us can read it at the same time?"

The next day I took Margarita Timofeyeva a second copy of *The Truth About Chernobyl* and went up to see Zalygin.

"We're putting your story in the No. 10 issue for 1988," said Zalygin as soon as I had entered his office. "It's really shaken me up. Two and a half years have passed since Chernobyl, but most of us still don't know anything about the disaster. We at *Novy Mir* are now in the curious situation of having two jobs to do: getting *The Truth About Chernobyl* and Solzhenitsyn's *Gulag Archipelago* into print.

"If the *Archipelago* is needed to strengthen and raise the people's spirits, *The Truth About Chernobyl* is needed to preserve life. We must figure out some way of getting a foreword from a distinguished nuclear expert. Given the political situation today, that's important. Censorship is still strong, as are the people behind it. I shall write the introduction myself."

At this point I found it necessary to calm Zalygin's ardor. I was most grateful to him, but my own harsh experience of battling the censors and the nuclear mafia weighed heavily on me. They were still out to get me, still hunting down my manuscript, and they were quite determined that it should not be printed.

"Sergei," I said. "I must warn you that it won't be easy to pull this off." I told him the whole story of my dealings with *Neva*, the military prosecutor, and Gosatom. "First and foremost, one thing has to be clear: we've got to bypass Gosatom and Glavlit at all costs, because they operate in tandem. We've got to move decisively in the direction of the Central Committee Politburo, better still to Aleksandr Yakovlev. He's someone who must support us. He's now in charge of their damned ideology. When all is said and done, the main censor in the country is the Central Committee of the party: that's a fact. We must go and see Yakovlev."

"I know him well," said Zalygin, agreeing with my point surprisingly quickly. "I'll go and see him tomorrow. But can't you see that there's no way we can bypass Glavlit? They have to sign approval for the entire issue to be printed."

"What if you don't let them have the issue for their signature, but merely publish it on behalf of the editors?" I asked naïvely.

"They'll stop the presses. They have the right to do so."

"But what about *perestroika*? How long will these bureaucratic watchdogs go on hiding the vandalism of the system from the public? What about the fresh winds of *perestroika*? Where are they? Sergei, you've got to bypass the lot of them."

"We have a very simple device that we use," said Zalygin. "If we're working with a particularly seditious story, we send it, say, to Gosatom, Minvodkhoz, or Goskomgidromet, and ask for their conclusions, which are usually negative. This doesn't bother us, though. We then assume the responsibility ourselves and go to print, and we often print the bone-crunching response of the government department right next to the story, together with our own comments."

"Even so, don't you need the Glavlit stamp?"

"Yes, we do. But it's a mere formality by then. We are

entitled to complain about them and fight, and go to the Central Committee. In most such cases we come to some agreement and the work is published."

"I still insist we bypass Gosatom."

"What if Glavlit doesn't give in? After all, they do have the blacklists from Gosatom, circulars, orders . . ."

"Shcherbina issued a secret order, which he himself signed, making it illegal for anyone to bypass Gosatom with Chernobyl-related material."

"Well, there you are then!"

"We've got to lodge a protest with the Central Committee against that order, at a press conference; in other words, we have to publicize it widely."

"Do you have the text of the order?" Zalygin inquired.

"I'll try to get it. In all probability it's very similar to the KGB order on Chernobyl, which says, 'It is illegal to publish for general circulation the real reasons for the Chernobyl accident . . .' Note how the KGB text calls it an accident, not a disaster. That one inaccuracy gives you some idea of how completely the state security people have the situation under control. Of course, it could have been a standard euphemism; they have always been fond of drooling euphemisms. But beneath it all they're cruel and murderous."

"All right then," said Zalygin. "In anything I do involving the publication of *The Truth About Chernobyl* I shall consult with you. We shall make the decisions together."

The apparent ease with which Zalygin agreed to carry out my plan, which must have seemed unrealistic at first, surprised me. But it also left me firmly convinced of his sincerity.

Aleksei Rakov phoned me and asked, "How are things going with *Neva*? Has Mr. Nikolsky printed your story?"

"No, he has not. He even gave the proofs to the military prosecutor. According to him, we didn't have the right stamp."

"Let me read it. You know you can take a plane from the Neva to the Don, and there is a journal called *Don*— are you familiar with it?"

"Yes, I am."

"And there's a fine Don Cossack called Vasily Voronov, the chief editor. Would you mind letting *Don* have 'The Reactor Unit'?"

"I'd be delighted."

"Bring the story along."

I went to *Druzhba* and handed Rakov "The Reactor Unit." Vladimir Firsov, the chief editor, called Rostov-na-Donu in my presence to speak to Vasily Voronov.

"Have you got the nuclear censors' stamp?"

"Yes, we have. The Minatomenergo stamp with the signature of Minister Reshetnikov."

"I hope you realize that without the nuclear stamp *Don* is never going to lay an egg like 'The Reactor Unit.' It just won't happen."

Firsov called to say that Voronov wanted the manuscript sent as soon as possible. He was going to put it in the first issue for 1989.

Exactly ten years after I wrote it, I thought. It shouldn't have taken that long!

*

Within a week Rakov had read "The Reactor Unit."

"What do you think?" I asked him.

"It's hair-raising stuff. And to think that was all before Chernobyl. Good God! It's really idiotic! Suicidal! It's got to be published. If it had been published before

Chernobyl, perhaps there never would have been a disaster at all."

This was another manifestation of a pious Russian belief in the omnipotence of the Word, something that I had encountered several times previously. The Word is God!

A people that believes so strongly in the Word is truly praiseworthy. But words have to be turned into deeds, as God commanded. Good words into Good deeds. And may Russia become ever stronger!

*

Practically every day I visited the editorial offices of *Novy Mir*, so as to keep track of developments and avoid mistakes. Margarita Timofeyeva and Inna Borisova had read *The Truth About Chernobyl*.

Timofeyeva said, "You have wounded me. I was in tears. I'm walking around with a tight feeling in my chest. It's terrible, what's happened to us, a tragedy. And people still haven't read it."

Borisova was more restrained. "I don't know how to put it. It's terrible," she said with an intense look in her eyes. "We must publish it in a single issue. If we put it in two, they may block the second one."

The board of editors decided to publish the entire work in a single issue. It would have to be shortened somewhat. Someone proposed a new title consisting of a pun on the name *Chernobyl*—namely, *Chornaya Byl* (*Black True Story*). This title was even used on the first set of proofs. But I insisted on keeping *Chernobyl Notebook*.*

---

* Title of the original Russian-language edition of *The Truth About Chernobyl*.

*
*

When he went to see Aleksandr Yakovlev, Zalygin was carrying a letter addressed to him. Just about that time, however, Yakovlev's role in the Politburo had shifted from ideology to international affairs. The ideological sector was now under Vadim Medvedev.

Zalygin told me that Yakovlev had read the letter and was favorably disposed toward the publication of *The Truth About Chernobyl*. Zalygin then went to see Vadim Medvedev, who despite his tremendous energy was still feeling his way in his new post. Zalygin handed him the letter and asked him to support publication. Medvedev suggested, "Sergei, I think you'd better take the manuscript to Gosatom. Let them turn it down, and then we will support the journal."

*
*

"No, Sergei," I said to Zalygin after I heard his account of the visit to Yakovlev and Medvedev. "We must not send this story to Gosatom. They will go behind our backs and arrange for the government, the Central Committee secretariat, and even the Politburo to ban publication. Nobody at the top is going to read the text. If they did, we might have had a favorable decision today. So far we have only one big advantage: we have the manuscript. Gosatom hates this manuscript. For the time being let it all be separate."

"What shall we do?" asked Zalygin.

"Wait till the right moment," I replied, "or at least try to reach Gorbachev."

"I'm not convinced that they will resolve it there. But I'll get in touch with Gorbachev's assistant and request a meeting with him," said Zalygin.

Meanwhile, Timofeyeva and I had shortened the story by two pages of printed text, and the editors had sent it off to be composed for inclusion in issue No. 10 of 1988. This process was completed quickly, and by September we had the proofs.

Then it occurred to us to ask Andrei Sakharov to write a foreword and give us his support. Natalia Dolotova, a senior editor in the *Novy Mir* prose section, took a copy of the proofs to Sakharov. Her late husband, also a physicist, had known Sakharov during a long career in the nuclear energy field. Dolotova's editorial colleagues felt that it would be easiest for her to approach the great man.

On 18 August 1988 Academician Sakharov read *The Truth About Chernobyl* in the *Novy Mir* version. Within three days he wrote a foreword, which was promptly sent to the printers and composed.

On 9 September there was a second set of proofs of *The Truth About Chernobyl*, which I read and checked for errors.

Zalygin told me that he had given his signed approval for the publication of issue No. 10 and had sent the text of the entire issue to Vladimir Solodin at Glavlit—the man who had conducted an investigation of "The Reactor Unit" when it appeared in *Neva*, and who together with Gosatom had given his blessing to the inquiries conducted by the military prosecutor. The outlook was not good.

After reading *The Truth About Chernobyl*, Solodin is reported to have said, "It's obvious to any halfwit that this story tells the unadulterated truth. That's why it's dangerous. But it would be immoral not to print it." A remarkable utterance, for which I was very grateful.

Solodin did not, however, give his signed approval for

publication. His argument was twofold: Gosatom had enacted a ban; and Gorbachev, while on a trip to India, had signed a contract with Rajiv Ghandi for the construction by the Soviet Union of a nuclear power station. For Glavlit, which put its own interpretation on each move of the top leadership, this meant that the general secretary endorsed nuclear power.

The story was taken out of issue No. 10 and moved into No. 11.

It is to Solodin's credit that, when returning the story without his signed approval for publication, he warned Zalygin and me that we should not under any circumstances send *The Truth About Chernobyl* to Gosatom.

This gesture marked a new departure in the conduct of Glavlit, which was clearly in conflict with the censors from government departments and anxious to break free of the prohibitions the censors sought to impose, leaving itself with the final say on such matters.

That, however, was the beginning of the end not only for censorship by government departments but also for Glavlit itself. Perhaps it was a kind of self-liquidation under the influence of the *aqua regis* of *perestroika*?

Being the best educated and most cultivated of the censors, Solodin understood that their departure should be started voluntarily, and thus with fewer moral and physical losses. Glavlit had already begun to prepare a letter to the Central Committee Politburo on the inadmissibility of censorship by government departments, since such censorship was designed to prevent the public from learning not only about state secrets but also about the failings—and, in some cases, crimes—in the work of the departments themselves.

**
*

Meanwhile, *The Truth About Chernobyl* was postponed to the No. 12 issue of 1988. Time was slipping through our fingers. Lies were still being propagated about Chernobyl. Shcherbina, Mayorets, Verkhovykh (Slavsky), and Marin (Central Committee) were awarded medals for their part in Chernobyl. Goskomgidromet (the State Committee on Hydrology and Meteorology) and Minzdrav (the Ministry of Public Health under Ilin) were following Shcherbina's orders to persuade the public that Chernobyl had been not a disaster but just a minor accident, and that it was possible and even advantageous to live on slightly contaminated land. There was much official praise for the clean-up personnel, who already numbered some six hundred thousand. In actual fact, the needs of this crew were completely forgotten: they were in effect written off, denied medical assistance, and accused of being malingerers.

There was reason to believe that Shcherbina and his team were unaware of the impending publication. Confident that Grigori Medvedev had been put to rest for good, they rested on their laurels.

During a bout of depression Sergei Zalygin once again considered sending *The Truth About Chernobyl* to Gosatom, right into the jaws of the beast.

"Out of the question! Forget it!" I fired back. "Believe me, Sergei, over the past ten years I have acquired enormous experience fighting the government departmental censors. They're just like a monster with a hundred heads."

"But what should we do?" Zalygin asked.

"Wait, and lie low." And then I remembered something. "I've just had an idea. I have to meet with Academician Sakharov to seek his advice and discuss certain matters. Perhaps I could ask him to write a letter to Gorbachev, jointly with you."

We agreed on this course of action. As requested by Zalygin, Dolotova arranged the meeting with Sakharov. I accompanied her on 18 October 1988.

Around 8:30 p.m. we arrived at Sakharov's apartment building on Chkalova, 68, near the Kursk railroad station. We took the elevator to the sixth floor. The door was opened by Yelena Bonner, a short, frail woman wearing thick glasses, her black hair heavily streaked with gray. She appeared slightly confused, saying that her husband was still asleep upstairs. Stairs leading to the second floor of the apartment were visible at the end of a long corridor off to the left.

Yelena invited us into the room nearest the entrance; against the wall just inside the door, on the right, was a computer and printer. She apologized as she sat down at the keyboard and finished typing what she had been working on. Then she said, "Why don't we go into the kitchen? There's more room in there. We can wait for Andrei in there. I'll make some tea."

We passed another room in which a color television was on and entered the kitchen.

I had brought along a copy of my book *A Moment of Life* that I had inscribed to Andrei Sakharov. I took it out of my briefcase, together with the proofs of *The Truth About Chernobyl*, which I intended to hand to Andrei as soon as he appeared. I put the book on the table next to the window.

There was an awkward silence. Yelena said that she and her husband had read *The Truth About Chernobyl* straight through in a single night and had found it deeply disturbing.

We heard footsteps on the stairs, and Andrei Sakharov soon appeared in the narrow corridor that led to the kitchen. He was wearing faded jeans and a white open-necked shirt. His jeans were slung rather low and rolled up slightly at the waist. He was very thin, bordering on emaciated. His face was the enlightened face of a great martyr, his clear, light blue eyes wholly trusting and childlike. The overall impression was that the man was a saint. He seemed to enter the kitchen cautiously, with a slightly unsteady gait. His appearance suggested an immense, otherworldly wisdom. His evident exhaustion showed that his sufferings and struggles over many years had cost him dearly.

He's a saint, I found myself thinking, and immediately decided that it would be presumptuous and brazen of me to make him a gift of my book, even though it too had been born of suffering.

With evident curiosity, Sakharov stared several times at the book, which lay next to the window. The atmosphere was slightly tense. Nobody said anything. Natalia was blushing. My heart was pounding, as if I were a first-year student at his first exam, and in the presence of the leading physicist of the day; it was practically like being in front of God himself. Suddenly I blurted out, "Andrei!" My voice sounded too resonant and loud, but I was really wound up and could not stop myself.

Sakharov looked at me sideways, with an encouraging smile at the corners of his mouth.

"Many thanks for reading *The Truth About Chernobyl* so quickly. It's a great honor for me. And thanks for the foreword."

Suddenly in his throaty alto voice, with a characteristic burr, he said, "Lelya and I read it in a single night. It really shook us up. It's hard to surprise me, but I hadn't thought things were quite that bad." Then he said quietly, "So how are things going? Is everything all right with the book?"

"That's just it, Andrei, it's not going well at all!" My confidence was restored by his startling ability to get straight to the point. As a great scientist, with little strength or time remaining, he knew how to avoid complicating issues and vague talk.

I briefly explained how Shcherbina's secret circular, now in the possession of Glavlit, required all material about Chernobyl to be sent to Gosatom. I also told him about Zalygin's trip to the Politburo to see Yakovlev and Medvedev, and how Medvedev wanted the story sent to Gosatom for their reaction, although he promised to offer his support for its publication even if they rejected it.

"Gosatom—you mean to Petrosian?" Sakharov asked.

"That's right."

"No. It mustn't go to Gosatom. But who *can* it go to? Perhaps to the Council of Ministers, to Shcherbina? Or to some institute. But the Kurchatov Institute of Atomic Energy is also out of the question."

His head bowed, Sakharov sat with his mobile and capable-looking hands on the table. His hands seemed to me almost capable of thought.

"What are we going to do?" he addressed the assembled company in his throaty alto voice.

"Shcherbina is out of the question," I said. "He's very much an interested party; he's afraid that the book will say he was opposed to evacuation. We've got to go to Gorbachev. He's the top censor in this country."

"I shall write a letter to Gorbachev," said Sakharov resolutely. "And let's do nothing with the manuscript for the time being. Give me a little time and then call me."

Yelena served tea, white bread, butter, Lithuanian

cheese, and curds. We talked about the situation in the country, about various journals—*Ogonyok*, *Nash Sovremennik*, and *Molodaya Gvardia*. Natalia Dolotova passed on to Sakharov Zalygin's invitation to become a member of the board of editors of *Novy Mir*. Andrei expressed gratitude and said that Vladimir Maksimov had already arranged for him to join the board of editors of *Kontinent* in Paris. He was glad to hear the news from *Novy Mir*. We discussed the need to nominate Andrei Sakharov as a candidate for the Soviet Congress of People's Deputies. When Yelena, in a light-hearted vein, said that he should become a candidate only if Gorbachev made him his first deputy, there was embarrassed laughter all around.

It occurred to me again that Andrei looked unwell. Every now and then he had a distant look in his eye, as if his thoughts had transported him into the most distant reaches of the heavens.

There was a phone call in the corridor for Andrei. He talked excitedly and at length about the problems of physics and field theory. He seemed quite a different person, full of zest and explosive energy, uttering a torrent of animated words about the most complex problems of theoretical physics. I found myself thinking that he was now soaring effortlessly in his true element; now that he was immersed in his life's work he seemed to have broken free of the oppressive atmosphere of the sociopolitical struggle, in which he had become involved largely at the expense of his beloved science. He had done so solely because, as a fearless and wise man, he really loved people.

We soon said good-bye. As they saw us out, Andrei and Yelena wished us well and complimented me again on *The Truth About Chernobyl*.

**
*

A while later, when *The Truth About Chernobyl* had been published, I asked Dolotova to send a copy of the book to Sakharov.

Sakharov kept his promise. On 10 November 1988 he wrote a letter to Mikhail Gorbachev. It read as follows:

To the Chairman of the Presidium of the Supreme Soviet of the USSR,
General Secretary of the CPSU,
Comrade M. S. Gorbachev.

Dear Mikhail Sergeyevich,

I wish to draw your attention to the fact that in our state the public, and public opinion, have not been allowed to become involved in problems pertaining to the planning and siting of nuclear power stations.

This will lead, and has in fact already led, to mistakes and grave complications, since the experts are unable, despite their good intentions, to take into account all economic, ecological, demographic, social, and other conditions and circumstances. This can be done only by that same public, among which one always finds specialists and experts in any field of knowledge, but which narrow specialists persist in describing as a collection of dilettantes. And this, despite the fact that shifting part of their own responsibility onto the public is precisely in the interests of those same experts.

All developed countries have long realized that public involvement in the solution of technical tasks is essential, and that both technical progress and the scientific and technical revolution thereby acquire a human dimension.

I felt that these general points needed to be made in connection with a concrete fact: the publication of Grigori Medvedev's story *The Truth About Chernobyl* in the journal *Novy Mir* has been delayed. (I am the author of the foreword to this story, and Mr. Medvedev is a nuclear engineer and a writer.)

I am convinced that the public not only can but must know all the circumstances of the Chernobyl disaster, despite the devious maneuvers, despite the interests and the ambitions of the censors in certain government departments.

In this matter any limitations are damaging to our society, to the memory of those who have lost their lives, and to the morale of those made sick as a result of Chernobyl. By hiding the facts we are enhancing the possibility of their recurrence. This cover-up is illogical, if only because all the facts have been reported by us to the IAEA. Does this mean that world public opinion is entitled to know more about events in the Soviet Union than we are?

And if we hide the names of those to blame for the disaster (and who are now engaged in censorship), does this mean that we are thereby providing full security for future bunglers in the planning and construction of nuclear power stations?

There are many such considerations which occur to me, though I shall not mention them all. As a scientist who has been directly involved in the problems of the utilization of nuclear energy, I feel that it is no longer possible for me to withhold from the world public my opinion of the work of Grigori Medvedev, and I intend to ensure that it is widely known.

It is my view that the publication of this story in *Novy Mir* would be of great service to our state; it would greatly broaden public discourse and self-knowledge on the part of the Soviet public, without whose involvement our further development is inconceivable.

With the most profound respect,
Academician Andrei D. Sakharov
Member of the Presidium of the USSR Academy of Sciences.
4 November 1988

Sakharov's letter to Gorbachev was mailed on 10 November 1988, and was received despite the possibility that it could have failed to reach him.

Zalygin soon received a phone call from an assistant to Mr. Gorbachev who said that Academician Sakharov's letter had been considered in the Politburo and that a decision had been reached: as the author was unwilling to send the manuscript of the story to Gosatom, he should send it to Vladimir Marin at the USSR Council of Ministers. It was understood that he would consider it and present a conclusion.

In response to a telegram from *Novy Mir* I went to see Zalygin, who was clearly encouraged by the new Central Committee decision.

"So shall we send the story to Marin?" he asked. "I was just waiting for your consent. I've already called him to say we'll be bringing the manuscript along shortly."

"Not to Marin, that's impossible," I said, well aware that this would upset Zalygin. "Don't forget that, after cutbacks at the Central Committee section on nuclear power, he switched to the Council of Ministers and is now Shcherbina's deputy. Giving it to Marin is like giving it to Shcherbina. That's all he's waiting for."

Zalygin clearly found my remarks disconcerting.

"What should we do then?"

"Say that the author does not agree, and leave it at that."

"Out of the question," said Zalygin calmly. "Sakharov's letter has been submitted to the Politburo for scrutiny. A day from now Gorbachev's assistant will be calling me and demanding that I send the manuscript to Marin."

"Tell the assistant that the author has taken the text to make some further changes. For a month. Sergei, we've got to lie low. Things are developing in the country at an astronomical speed these days; within a month a lot will have changed."

To my immense satisfaction, Zalygin found his way to agreeing with my arguments.

"It's too bad they didn't read the proofs of the story in the Politburo," I said. "They would have sent it straight out for composition, without going through Marin at all, and they would have signed an authorization to publish without going through Glavlit."

"But how can you get them to read it?" asked Zalygin. "I offered it to Medvedev and Yakovlev, but they wouldn't take it."

"Maybe something will turn up."

On that note we parted. And tentatively transferred the proofs of *The Truth About Chernobyl* to the second half of 1989.

*

Something did turn up. On 18 December 1988, when I went to the offices of *Novy Mir* to find out whether there was any news, Igor Dedkov, a critic, happened to be there. We were introduced.

"I've already heard of *The Truth About Chernobyl*," said Igor. "The proofs are being circulated around Moscow. It's aroused a lot of interest. It hasn't reached me yet. Do you have a spare set of proofs?"

Fortunately I did—a set that I kept with me, just in case they were needed. I handed it to him, without any particular reason to feel hopeful.

"Call me in a week," said Dedkov, who worked for the journal *Kommunist*. "If we're in luck," he said, "we'll print a bit of it in *Kommunist,* and that will help get the whole thing published."

"Here's hoping!"

Dedkov had finished reading *The Truth About Chernobyl* by the time I called him back on 27 December 1988. He said, "I'm overwhelmed. I've had chest pains for the past two days; I've been taking cardiac medicine. It's just perfect. I gave the proofs to Bikkenin, the chief editor, who's very close to Gorbachev. Be patient now. . . ."

***

Meanwhile, the journal *Don,* in Rostov-na-Donu, had at long last managed to arrive at the brink of publishing "The Reactor Unit," which would appear in issue No. 1 for 1989, ten years after it was written. Better late than never. And it's an ill wind that blows nobody any good: the ecologists of the Rostov area later used "The Reactor Unit" in their just struggle to oppose the construction and start-up of the Rostov nuclear power station. It's gratifying that they eventually won, and that the station was converted to the thermal mode.

I am deeply grateful to the chief editor of *Don,* Vasily Voronov, the valiant Don Cossack who succeeded in helping the story reach its readers.

On 2 January 1989 I called Dedkov at home. He sounded excited.

"Bikkenin read *The Truth About Chernobyl* really fast. He wants to publish parts of it in *Kommunist.* He passed it on to Frolov, who is an assistant to Gorbachev; that means the whole Politburo will read it."

Meanwhile, the public onslaught on censorship continued.

On 18 January 1989, in his speech to the Plenum of the USSR Writers' Union, Zalygin mentioned *The Truth About Chernobyl* and said that he considered its publication a vital necessity. He also delivered a stinging attack on government departmental censorship.

During a televised roundtable discussion among the chief editors of the principal literary journals on 4 February, Zalygin's deputy, Vladimir Kostrov, spoke passionately about *The Truth About Chernobyl*. At the same time he once again blasted the censors for keeping such a crucial work hidden from the public.

The journal *Kommunist* was preparing a major blow against censorship and Gosatom's entire system of coverup and prohibitions. On 3 March 1989 *Kommunist* published fragments of *The Truth About Chernobyl* under the title "Incompetence." There could be no doubt that *Kommunist* went ahead and published these excerpts precisely because it knew that Zalygin was going to publish the whole work. In fact, they used *Novy Mir*'s proofs for the purpose. This publishing event sent shock waves through the camp of those opposed to the truth about Chernobyl. At the same time it aroused bitterness and hope among the population of the regions contaminated by radiation: bitterness because they had discovered the truth; hope because the necessary steps would now at long last be taken and it would be possible to speak truthfully about the disaster.

All the central and regional Byelorussian newspapers reprinted the article from *Kommunist*. Letters to the editor poured in. Finally, on 17 April 1989, Yu. A. Izrael, chairman of Goskomgidromet, published the first map of the contaminated areas of Ukraine, Byelorussia, and Russia. He, too, had waited until *Kommunist* took the initiative.

He had waited three years since the Chernobyl explosion. Immediately after the disaster, it would appear, he had been unable to tell the truth; he had lied while people were getting sick and absorbing their fateful doses of radiation.

Zalygin made his displeasure known to me in the most forceful terms. He accused *Kommunist* of skimming off the cream and dislodging *Novy Mir* from its prime position.

I disagreed. "Sergei, *Kommunist* helped *Novy Mir* in this instance, and *Novy Mir* helped *Kommunist*. Glavlit will not have the nerve now to refuse publication—which issue of *Novy Mir* were you planning to put *The Truth About Chernobyl* in? No. 6? Well, then they'll give their signed approval for publication in No. 6. There's no way they can back out now!"

I was proved right. Solodin, the deputy head of Glavlit, rushed to the editorial offices of *Kommunist* to clarify the general situation with regard to censorship. Glavlit had by now realized that publication of such material in an organ of the Central Committee meant that something was wrong, something had changed.

Aleksei Antipov, head of the science and education section, who had himself prepared the publication of "Incompetence," told me that Solodin had addressed the board of editors and announced that *The Truth About Chernobyl* would be published in its entirety, with forewords by Zalygin and Sakharov.

On 27 March 1989, only twenty-four days after the publication of fragments of *The Truth About Chernobyl* in *Kommunist*, I was summoned to Glavlit. Before I set out on my way to Kitaisky Proyezd, 7, Vidrashka, who was Zalygin's deputy, triumphantly declared, "A pass has already been ordered for you. And remember that this is only the second time in the whole history of *Novy Mir* that a writer has been summoned to Glavlit in person to defend his works. You should be proud of yourself!"

"All I want is to make sure that there will never be a third time, and that Glavlit will vanish from the face of Russia."

So you see, Lilia Khokhlova, although you signed a pledge not to divulge the working methods of Soviet censorship, I shall divulge them anyway, for the good of the cause. Of course, it's easier for me, as I did not sign such a pledge. I intend to make this information known so that the day will soon come when facts that in any case should not be kept secret are public knowledge.

On 27 March 1989 I went to Kitaisky Proyezd, 7, to the Minenergo building, in which I had worked for eight years, and proceeded to the sixth floor, where Glavlit had its offices. I could see a massive locked door at the end of a very long corridor. Beyond it lay Minenergo and Glavstroy. My old office was to the left of that door. While I had worked there, I had often tried to figure out ways of getting the better of the monster Glavlit. And now . . .

I entered the office of Sergei Savinich, the censor who had been instructed by Solodin to hear my defense of *The Truth About Chernobyl*. He was a good-looking young man of no more than thirty-five, with black hair and eyebrows, and light blue eyes. OK, I'm ready for you, I thought.

The proofs of my story lay on his desk, with nasty-looking red pencil marks on practically every page. And there were questions. "Where did you get this name? Has it been published anywhere in the press? Where did you get this dose from? Who mentioned the name of this general? And who published these statistics on radioactivity?" And so on.

I was the one who had written *The Truth About Chernobyl*, and no one else. And for the past three years I had kept the Chernobyl file with me, together with the knowledge gained through my experience as a nuclear power expert.

I responded to all of Savinich's questions in two hours. While we were talking, Solodin came in. It was the first time I had ever looked him in the eye.

Do you think that you are the only one anxious to protect Russia, Vladimir? I asked him silently. What about me, and Zalygin, Dedkov, Antipov, Bikkenin, and millions of ordinary people in Russia, the Ukraine, and Byelorussia—indeed, people everywhere. Because now we are living not only in a Chernobyl state but on a Chernobyl planet.

On 28 April 1989 Solodin personally applied the authorization stamp of Glavlit to the proofs of my story. It read: "Permission granted for general publication."

On 5 June *Novy Mir*'s No. 6 issue for 1989, with *The Truth About Chernobyl* as its principal feature, entered circulation. At last the wall of nuclear secrecy surrounding Chernobyl, which had been erected so laboriously and unscrupulously by Gosatom and later by Shcherbina, had been breached. That wall of death now began to crumble.

Hundreds—thousands—of readers' letters poured in. More importantly, a tidal wave of truth about Chernobyl went crashing through the gap in the Gosatom wall and into all the mass media. That marked the beginning of the end of censorship by government departments, and of

Glavlit as well. Within a year the directorate was to disappear altogether.

But that is not all, Lilia Khokhlova: by way of a grand, definitive conclusion let me tell you another interesting and edifying story.

It happened on 12 January 1990, when around nine in the evening there came a knock at my front door. I opened to find a tall red-faced man with graying temples; he was wearing a gray hat and a long, slightly tattered leather overcoat. He carried a black briefcase. I could not understand what my visitor was mumbling, but he was clearly agitated. With trembling hands he produced some kind of identity document, tried to show it to me, and continued saying something about *Novy Mir* and Zalygin.

"Who are you and where did you come from?" I asked him, realizing that it is easier to answer questions than to put one's own thoughts into words extemporaneously.

"I'm from Novosibirsk," he said cheerfully.

"Where do you work?"

"In the Novosibirsk division of the Academy of Sciences."

"Who gave you my address?"

"I went to see Zalygin at *Novy Mir*. We talked for two hours," said the man in the leather coat, now speaking more confidently. "He advised me to get in touch with you. Later on we should both go and see him."

"What's your name?"

"Lev Maksamov."

Zalygin? I thought. That's strange. This fellow doesn't seem to be making sense. I've got to check to see if he's telling the truth.

In any case I decided to let him into the corridor, where I could question him more thoroughly.

"What have you come about?"

"Chernobyl. It's very important. I read your story and was very shocked by it. People are saying . . . "

"But who was it that gave you my home address?"

"It took me ages to find you. I got it through Minsredmash, through Kruglov, the head of the scientific and technical division at Minsredmash."

"Kruglov . . . " I cast my mind back but was unable to remember any Kruglov, although the name seemed familiar.

"Fine. So what do you want?"

"I want to talk to you. It'll take about two hours."

"I haven't got two hours. I don't receive visitors at home, and I don't talk at home to people I have not invited. I'll see you tomorrow at midday, at *Novy Mir*, we can talk there."

Evidently embarrassed, he opened his briefcase slightly and then closed it again; all the while he supported it in one knee. He smelled of liquor.

Trying to appear tough, I thought. Or maybe this was an old habit of people who worked for Minsredmash.

The man had dark, sinister eyes. His entire manner and tone were becoming increasingly insolent. He was the kind of person I instinctively dislike.

"Tomorrow at midday, at *Novy Mir*," I repeated curtly. "Good-bye!"

My visitor turned to leave, but reluctantly.

"Do you know how to get to the subway platform?" I asked him. "If not, I can take you there."

"I have a car."

I went to the kitchen and looked out the window to see what kind of car: it was a black Volga with a radio antenna.

So that's it! He obviously has a transmitter in his brief-case. Right, we'll see.

*

A year and a half had now elapsed since the publication of *The Truth About Chernobyl*, but my enemies were still trying to figure out a way to catch me off guard. Shcherbina and Mayorets were no longer in office. After *The Truth About Chernobyl* there was simply no place for them anywhere in government; and Ryzhkov did not include them.

A whole wave of disclosures had followed. Without extraordinary haste, literally everything was being opened to public scrutiny—even nuclear weapons reactors, the mere mention of whose names would in the past have landed one in jail. The newspapers were now vying with one another to discuss the 1957 disaster at Chelyabinsk. I had already written about that, though not much earlier, it is true. But I did write about it. "The Reactor Unit" was published in *Don* in January 1989, whereas the initial press coverage of the subject did not appear until September of that same year, on the thirty-second anniversary of the explosion. Of course, if "The Reactor Unit" had been published, say, in *Novy Mir* under Narovcha-tov, the subject could have been aired as early as 1979. It really is possible that Chernobyl might have been avoided. *Novy Mir* is taken very seriously and is very widely read. Its coverage of this issue could not have been ignored. And nuclear power operators, after reading the story, would have had plenty to think about.

But no! It seems that, through the censorship it had nurtured, the administrative-command system almost wanted Chernobyl to happen. Its wishes were granted.

What was the present objective of Lev Maksamov and his masters? And who were his masters?

*
**

I arrived at the *Novy Mir* offices around midday on 13 January 1990. Maksamov showed up half an hour later, in the same Volga with the radio antenna. The editor on duty was Nina Ivanovna. Zalygin was abroad on business until 22 January. Feodosy Vidrashka was not yet in the office.

Maksamov's eyes had a brazen glint in them; he seemed vigorous and determined to fulfill the task assigned to him.

Let's see what kind of lion this Lev is, I thought. "Perhaps Lev and I could talk in Sergei's office?" I asked Nina Ivanovna.

She told us we could, and so the two of us entered the room that had played such an important role in Russian literature. A single glance at the cocky, brazen expression on Maksamov's face might have justified dismissing him then and there, but Zalygin had asked me to meet him. And it was obvious that there was someone else behind Maksamov—but who?

That morning, before going to *Novy Mir*, I had conducted a background check on Maksamov. I had called the Council of Ministers and talked with people I knew there. No one had ever heard of him. As Maksamov had mentioned Kruglov, I called him at Minsredmash. With an awkward laugh Kruglov said that he did know the man. He had worked for fifteen years at Minsredmash, at the Novosibirsk plant that produced nuclear fuel for nuclear power stations. Then, having specialized in powder metallurgy, he switched to the steel and alloy institute of the Novosibirsk branch of the Soviet Academy of Sciences. After a disagreement with the management he had been unemployed for eight months. He was a purveyor of the Yid-Mason version of the Chernobyl disaster. I now began to understand. I felt I should listen to him; it might even be interesting.

Maksamov and I sat facing each other in armchairs near the chief editor's desk. He opened his briefcase and left the lid in the vertical position.

An antenna, I thought.

The contents of the briefcase were covered with a thin layer of business papers, so that I got the impression they must be resting on something hard with a small gap.

"Tell me about yourself," I said to Maksamov, looking him in the eye. He had a purple, somewhat lined face with high cheekbones; his thick, wavy hair, no doubt jet black in his younger days, was entirely gray. He had a stocky, athletic build. He had dark eyes of a slightly Mongol type. Maksamov was tense and sweating.

"I am a physicist, with a graduate degree in physics and mathematics. I worked for fifteen years at the Novosibirsk factory that produces fuel assemblies for nuclear power stations. I did pretty well there, and then I moved to the steel and alloy institute of the Novosibirsk branch of the Academy of Sciences. That's positively crawling with Yids. It's not that I am against Jews. There are some Russians who are worse than Jews. And some Jews—for example, Budker—are better than some other Russians. In fact, I would say that Academician Budker is Russian and that Ivanov is a Yid, even though he is ethnically a Russian. The Zionists gave that Budker such a hard time that he refused to work for them. I was afraid they would finish me off, too. That's the only way they know to deal with people like me. Anyway, Grigori, I have read *The Truth About Chernobyl*. It upset me tremendously. There's never been a book like that about Chernobyl, and there never will be, I'm sure. What struck me most about 'The Reactor Unit' was the calm, objective tone with which events are described. It's really sincere and objective. I was wondering if you might wish to continue the story. There are a number of points that could be enlarged on. Which practically brings us to the main subject of our conversation—then you will see what I mean."

"What did you discuss with Zalygin?" I asked.

"The same thing we are about to discuss," said Maksamov in a sinister tone of voice. "When he heard what I had to say, he asked me to get in touch with you, so that we could both go and see him. Zalygin and I agreed that we would record our entire conversation."

As we sat down at the news desk Maksamov became animated, his eyes glinting with anticipation. Clearly everything was going as he had planned. His open briefcase containing the two-way radio was on the desk.

They're going to steer the conversation, I thought. He'll soon be running to the phone to consult.

"Well, let's begin then," said Maksamov with an inquisitive stare, as if he were sizing me up. "When the Chernobyl explosion happened, I, and with my assistance the Novosibirsk KGB, pointed out that in our young people's newspaper, *Sem Dnyei*"—he pulled out an original of the 15 April 1986 issue, exactly ten days before the disaster—"here, in the lower right-hand corner on the back page, beneath the movie, theater, radio, and TV programs, they had all of a sudden put a stylized diagram of an RBMK reactor: a double square, with the circles for the fuel channels inside." Maksamov produced a photo-copy of a page from an encyclopedia with the same emblem under the heading "Uranium-graphite reactor." "Now if you place a Jewish calendar over the newspaper page and read from right to left, starting at the space corresponding to 15 April in the Gregorian calendar, when you come to 25 April, you see a picture of a nuclear explosion in the shape of an irregular six-pointed star of David. The top left and right points resemble the hands of a clock, don't they? The time they show is between one and two in the morning—in other words, the night of 26 April. Now if you hold the paper up to the light, just opposite the symbol of the explosion are the letters YCCP, which stand for the Ukrainian Soviet Socialist Republic. That's the site of the accident. Got it?" said Maksamov excitedly. "The Ukraine is mentioned so that all the Jews could be warned and could escape in time."

"As far as I am aware," I said, "the uncontrolled exodus from Kiev did not begin until after the explosion."

"Ah, but you don't know everything," said Maksamov with a frown.

"And what is it that you know?" I inquired in turn. "I warn you that you must tell me everything about this question; otherwise our conversation must end right here and now."

"Oh, very well then," said Maksamov, capitulating.

I imagined that there would soon be a phone call for him, but I was wrong: he himself placed the call. He had to say where he was and give his handlers the phone number.

"Just a minute, I have to call Aleksandr Nikolayevich. We had an arrangement."

That's exactly right, I thought, you certainly did have an arrangement. "And who is Aleksandr Nikolayevich?" I asked.

"Tikhomirov, do you know him? He's a well-known TV journalist and moderator, with a dark complexion and a big face."

Could Tikhomirov be directing our conversation? I wondered. Surely it must be some KGB general, and the name *Tikhomirov* is a cover. All right, then, it's Tikhomirov. It's hard to tell them apart and to know who's a KGB general, who's a TV reporter, and who's working for Novosti.

"Yes, we're working, Aleksandr Nikolayevich."

Aleksandr Nikolayevich could, for example, be Deputy Minister Uzunov, in Minsredmash. But there are plenty of Aleksandr Nikolayeviches around. Let's get on with it!

Maksamov, who had by now forgotten the rules of the game and allowed his mask to slip, listened to his instructions with a servile look on his face, as if mentally standing to attention. Werewolf . . .

"Yes, I understand. We're in Zalygin's office, there's a phone," he stammered into the receiver. I found myself thinking of *Literaturnaya Uchoba*, of Teplukhin—he used to get instructions over the phone, too.

Maksamov came back to where I was sitting.

"Where were we? Oh, yes, the time and place. . . . That's really funny: it's like something out of a novel by Yuri Trifonov. . . . Right."

"Did anyone ever find out how the diagram of a reactor found its way into *Sem Dnyei*?" I asked.

"Yes, they did," Maksamov replied reluctantly. He had not been expecting questions like these, and he clearly found them disconcerting.

185

The way these people work is so dumb! That's why we're losing all over the place, I thought.

"So where did they come from?"

"We'll come back to that."

"No, I'd like you to tell me."

"It's the emblem of the Institute of Nuclear Physics of the Novosibirsk branch of the Academy of Sciences."

"And were there any other emblems—that is, emblems of other institutes belonging to that same branch of the Academy?"

"Yes, there were."

"Do you have them with you?"

"Here they are," he said, showing me several other emblems of various institutes of the Novosibirsk branch.

"That means," I said, trying to summarize, "that the editors, at the request of a branch of the Academy of Sciences, had to publish all the emblems it had received? Were the clichés ready?"

"Yes."

"So it's not surprising that the first one to be published was the emblem of the Institute of Nuclear Physics," I said.

"You really don't know the Jews!" said Maksamov heatedly. "They had prepared all the emblems precisely so as to print this one in particular and pass on information."

"So be it. What next?"

"Now take a look at this," he said, producing photocopies of *Pravda* for 1 and 2 May 1986, which contained coverage of the May Day celebrations in all major towns and all the republics in the country. "This is curious: near the headline of the story from Kiev there are two carnations instead of the normal three. The question is: who stands to gain by making a mockery of the Slavs—Ukrainians, Byelorussians, and Russians—and by insinuating in many millions of newspapers that the land in which we live has now become the graveyard of the nation?"

"What about the graveyard built in the years before Chernobyl by Father Stalin and his role model, Lenin, with a hundred and ten million bodies in it?"

From his expression I could tell that Maksamov did not like my question.

"That, too, naturally, but Chernobyl was much worse."

"Chernobyl was the continuation of genocide by means of radiation. And it became genocide because people lied and constantly played down the dangers."

"The Jews are the only ones who stand to gain from all that. It fits the goals of the Yid-Mason conspiracy."

"Does that mean you dispute the version in *The Truth About Chernobyl*?" I asked.

"Not at all!" Maksamov replied emphatically. "It's just that point in the story where Toptunov and Akimov raised the rods in an attempt to increase the power of the reactor. Obviously they should not have done that. Why was the operational reactivity reserve so low at the time? Were the operators really that stupid? There could have been sabotage: someone had raised the rods some time before, controlling them from backup control panels in the nonoperational rooms, and the people working inside the unit only made a bad situation worse by raising the rods. Could that have happened?"

"Let's assume that such evil people existed. First of all, they would have had to be very well educated, and with a thorough knowledge of the controls of the emergency protection system. Also, they would have had to be consumed by visceral hatred of our country, its culture, and its history. Third—and most important—the backup control rooms and closets were locked and sealed while the unit was in operation. Keys were given out only in special cases; and if any of the closets were opened at all, they were opened in the presence of the shift foreman and had to be recorded in the log."

"With the kind of mess we're used to, anything is possible," Maksamov countered. His facial expression was now businesslike and even constructive; he clearly liked my line of reasoning. "And remember, Russia has plenty of enemies, even counting the Yid-Masons alone."

"However, your version of events would have to be checked out and proved," I said. "Who was at the control panel during the experiment, or a long time before it? Was this reported in any documents or in the press?"

"Yes, it was!" said Maksamov with delight as he reached for a clipping from *Pravda*, an article by Vlasin about accidents at nuclear power stations. "Apparently, a long time before the accident two scientists with advanced degrees proposed to collect all the control software used at nuclear power stations so as to study its shortcomings and, in the long run, increase efficiency. Assuming these two had malicious intentions, they could have found the commands by studying the software. And they might not have been able to foresee the consequences," said Maksamov in conclusion.

"That's a bit too ingenious," I said. "And highly speculative. But since the question has been raised, let the KGB conduct an investigation, with the help of the institutes concerned. That shouldn't be difficult."

"Now we've come to the final part of our conversation," said Maksamov gleefully.

"Not yet," I said. The phone rang. As Maksamov rushed to pick it up, it occurred to me that Tikhomirov clearly had other ideas.

"Yes, I understand. Of course."

"Tell me," he said, returning to the news desk. "What in particular do you find wrong with the Novosibirsk version of the accident?"

"It hasn't been proved. It would have to be examined thoroughly, and precise evidence would have to be presented. Otherwise . . . " At this point I decided to let Maksamov have a piece of news he was not expecting. "You know something, Mr. Maksamov?" I said, watching him closely. "I've known all about the Jewish version since 1986."

"How?" Maksamov grew pale but did not lose his composure, even though he had clearly fallen into what nuclear engineers call an iodine well. Leaning forward suddenly, he said in a low voice, "Where did you get this material?"

"I found out about it in one of the journals of the Komsomol. But the material itself I got from the KGB, from Viktor Chebrikov."

"Chebrikov is a Jew!" Maksamov exclaimed.

"What do you mean, a Jew? He's Russian!" I objected.

"His wife is Jewish."

"And Sakharov?" I said, pressing my point.

"He's a Jew. Sakharov is a Jew."

"What kind of Jew is he? He's 100 percent Russian. He has the Russian character, and Russian ancestry."

"He is Zuckerman, and not Sakharov. And his wife is a Yid. That's why Chebrikov stopped our material from getting through. We sent it to him from Novosibirsk, way back. He held it up deliberately. They're dead against *perestroika*. When Chernobyl blew up they called Gorbachev: 'So you're planning some more *perestroika*, are you, Misha? Just watch out. . . .' Then there was that explosion in Arzamas. And another phone call: 'Look, Misha, we warned you. . . .' Do you understand what is going on? This brazen gang is talking openly. But now I have reported this material to the people's deputies from Novosibirsk. With some help from me, the most aggressive one among them wrote a detailed report to Gorbachev. Here is the letter, with Gorbachev's decision. Take a look." Maksamov showed me the text of a letter; the decision was written at an angle across the top of the page. "That's not Gorbachev's handwriting," he hastened to add. "I transcribed his decision from the original copy. Here it says that N. Slyunkov was to be asked to conduct an inquiry. You see?" He thrust the letter toward me.

"I'm not interested in a decision that's not genuine," I said.

"But that's the exact Gorbachev text, I transcribed it to my copy. Here. Now I am working with Slyunkov. At the Central Committee I reported it all to Slyunkov. He could hardly believe his ears."

"Did he believe you?" I asked.

"Of course! He asked me how such a thing could happen. And I told him, 'That's the Jews for you!' Slyunkov asked me to prepare a list of members for the commission of inquiry. I included you. There's also a KGB general on the list; in fact, I drove with him yesterday in his car when I came to see you."

I know, dumbo, I said to myself.

"I wanted to include Velikhov, but I changed my mind.

I get the impression that he, too, has Yiddish sympathies. We included Laverov, an academician, and many other people, such as computer experts . . . and you."

"And what is this commission going to do?"

"It will get to the bottom of the whole thing, from the very beginning. We will analyze all the papers, diagrams, and records. We will question lots of people. But they could lie. So you, as an expert on nuclear power plant operations, will tell us whether someone is lying, and we will have his story checked again. Of course, the checking will be done behind everyone's backs."

They're trying to turn me into a stool pigeon, the rats! On the other hand, maybe this is a trap. "And who authorized you to include me as a member of the commission?" I asked Maksamov.

"Well, I thought you would agree . . . on something like this."

"It's certainly interesting. But I must ask you to drop my name, and right away! Call Slyunkov and have him delete my name from his copy."

"You want me to do that to a member of the Politburo?!"

"That's right! A member of the Politburo!"

Maksamov blushed a deep red.

I was convinced the phone was about to ring, but it did not. Maksamov himself rose and went to the phone. After receiving instructions from Tikhomirov, he sat down again and smiled.

"I've been thinking, Grigori Ustinovich. How is it that you don't seem scared? I, for one, am really afraid of the Zionists. They'll kill, the devils!"

"I'm not afraid of dying, so let them. What good would

it do them anyway? Now the people can speak; that's the main thing."

I wondered whether Tikhomirov was trying to intimidate me.

"But if you worked on the commission you could write a new, more exciting story. Just imagine the kind of material you'd have at your disposal!"

"Let Gubarev or Shcherbak write it."

"Gubarev won't do it. And I've already spoken to Shcherbak, he's really shaken up. He says he should be dismissed. He, too, is obviously afraid of the Zionists. Apart from that, quite frankly, some comrades, for example Marin, are saying some pretty nasty things about you. Particularly Marin. He's in a really foul mood."

"That's fine with me!"

"But when they see you on the commission run by Slyunkov, who is a member of the Politburo, they'll bite their tongues."

The way these people operate is really dumb, I thought wearily.

Maksamov now began to boast. He picked up another sheet of paper, actually a USSR government form on which a text had been printed. It was the decision to establish a special institute within the Novosibirsk branch of the Academy of Sciences, with Maksamov as its director.

This opened up the possibility that Maksamov was pushing the conspiracy theory as a way of returning a favor. Or maybe his handlers needed it as an additional weapon. Well, to hell with them anyway!

"Right, then!" I got up and, pacing up and down the office, said a second time, "Drop my name from that list. Go someplace else to get yourself a writer for your story.

There are plenty of them around these days, practically on every street corner."

Nina Ivanovna entered the room nervously and inquired whether we would be finished soon.

"We'll be leaving right away," I said.

Feodosy Vidrashka, finding on his arrival at work that Medvedev and Maksamov were meeting in chief editor Zalygin's office, had apparently flown into a rage and scolded poor Nina Ivanovna for letting us in there.

He was, however, afraid to come in himself to expel us—perhaps he was nervous about Maksamov. When we came out, he confronted Maksamov, who just stood there in an insolent, brazen pose.

"You! What right do you have to go into the chief editor's office?"

"Actually it was me, Feodosy, I'm to blame," I owned up to Vidrashka.

"That doesn't matter, you're one of us. As for comrade Maksamov here, when he came to see Zalygin I told him Sergei was busy. But he went ahead and just shoved his way into the office. He said he would be just five minutes, but he took up two hours of Sergei's time."

Maksamov looked condescendingly at Vidrashka and laughed.

That was all. In the corridor Maksamov again asked me to think the situation over. He said he would be in Novosibirsk until 25 January, and he asked me to call him on his return.

"Fine, get going! And good-bye!"

He realized that this was our last conversation and that I was standing firm, yet he repeated his request over and over again. What a bore!

Vidrashka was pacing around the lobby in a state of extreme agitation.

"That man, Grigori, is just plain bad news. You can see it in his eyes!"

Good riddance!

**
*

The next day I received a telegram from *Novy Mir* asking me to come as soon as possible. When I got there, Nina Ivanovna handed me a note that read, "Medvedev, author of *The Truth About Chernobyl*, has a meeting with Slyunkov on Friday at 2:30 p.m. Babanin."

"Do you know who Babanin is?" I asked Nina Ivanovna.

"Slyunkov's assistant. Here's his phone number."

When I called Babanin, I heard on the other end of the line the lordly voice of an old-time party apparatchik—the kind whose jaw sags as he talks to you, under the weight of his own importance.

"Good morning. My name is Medvedev. I got your note at *Novy Mir*. What am I needed for at a meeting with Slyunkov?"

"Yes," he said, his jaw sagging solemnly. "Actually it won't take long, perhaps a single meeting might suffice."

"Will you be elaborating the Jewish version put forward by Maksamov?"

"Yes, ha ha!" he said, slightly disconcerted but still solemn. "In a sense, yes. You know, there's nothing in your story about that version."

"My story deals with the substance of the question."

"Yes, I've read it. It's good. . . . So we'll be expecting you then."

"I'll have to think about it. I'll call you tomorrow to let you know what I've decided."

"Please, not later than eleven."

The next day I called Babanin at the appointed hour.

"This is Medvedev. Can you hear me? I regret that I will be unable to attend the meeting. The subject to be discussed falls outside my field of expertise. Good-bye!"

"What!?" Threatening grunts came from the other end of the line, but I hung up.

**
*

We shall assume that Maksamov and Babanin were the last gasp of a dying system of censorship, and of the old party apparat, now clearly on the verge of extinction. It is even possible that the "Jewish card" might have been a necessary weapon in the struggle for power within the party. Sort it out among yourselves, gentlemen! Do not drag literature into your sordid affairs. It will come along in due course to set the record straight.

So that, Lilia Khokhlova, my dear former censor, is that. It is just as well that Lilia was a former censor, that she left of her own accord, having become aware of the profound immorality of the monstrous dragon that for decades devoured the truth.

It is particularly sad to think that the people hired by the system to fill the ranks of its sinister corps of censors were mainly women. In this way, the system used the hands of mothers to choke the flow of spiritual oxygen to their own children.

I may not have given a complete description of the workings of Soviet censorship, but I am not a former censor, merely a disenfranchised Soviet writer. Nowadays there is a whole army of former censors. They should write their memoirs and open our eyes to the truth, telling us how they stifled the culture of their own country. There is nothing to stop them now: censorship ceased to exist on 1 August 1990.

Kashira, 17 August 1990

Postscript

AFTER COMPLETING MY STORY I thought, "Is there any point in writing this? Censorship is now finished."

Censorship is not really finished, however. On 16 January 1991, at the fourth session of the Supreme Soviet of the Union, President Gorbachev unexpectedly proposed suspending the nation's press law. What for? Admittedly, given the hostile reaction of the deputies, Gorbachev said that he was not insisting on the suspension of the press law but merely calling for objectivity on the part of the mass media. But even so . . .

The sinister shadow of Soviet censorship had once again loomed on the horizon. I was evidently too quick to send the censors off on vacation to write their memoirs. They will never write them, as they have signed a pledge of secrecy. But it would be most helpful for writers who have been battered by our censorship to describe its methods in each individual instance, thus gradually building up an impressive picture of the way a nation's culture was strangled. Once the writers have done this, it will be harder for the censors ever to come back into their own.

Moscow, 16 January 1991

Index

Accidents: in bacteriological weapons plant, 45–46; in coal industry, 9

Accidents, in nuclear power stations, 1, 2–3; as concealed, 9–10, 11, 13; Kamenev on, 99; KGB response to, 90–91; need to write about after Chernobyl disaster, 108–9, 111–12, 129–30, 147–48, 149, 177; need to write about before Chernobyl disaster, 39–40, 47, 49, 55, 60–61, 66, 81, 82, 87, 92, 100, 101, 102, 129–30; in post-Soviet regime, 2, 26; secrecy on, 40, 44, 47, 71–72, 78, 80, 82, 83–84. *See also* Censorship; Chernobyl nuclear power station, disaster at; *under specific nuclear power stations*

Administrative-command system, 130, 141, 180

Afghanistan, invasion of, 6, 56, 72, 73

Akimka, 65

Akimov, Aleksandr, 117, 121, 187

Akimova, Lyubov, 121, 123, 126, 127

Aleksandrov, A. P., 81

Alekseyev, Mikhail, 117–18

Alienation, reaction to Chernobyl causing, 19–20

Altaisky collective farm, 65

Andropov, Yuri, 14, 56, 72–73, 75, 82, 84

Anemias, Chernobyl disaster and, 24

Antipov, Aleksei, 175, 177

Aral Sea, 36

Arbatov, Georgii, 6

Arkhipina, Ludmila, 120, 121

Armyanskaya nuclear power station, accident at, 90

Artyushina, Valentina, 101, 109

Astafyev, Irina, 62, 63–64, 65

Astafyev, Maria, 62, 65, 70

Astafyev, Viktor, 43, 55, 57, 58, 61–71, 76, 77, 78, 79, 87, 101

ATETs (nuclear power and heating stations), 13–14

Atom. *See* Peaceful atom; Radioactive fallout

Atommash factory, 12

At the Last Boundary (Pikul), 58

Babanin, 194–95

Barkov, Viktor, 73–76, 90–91

Baruzdin, S. A., 82, 83, 84–85, 95

Batalin, Yuri, 118, 120

Belarus, 2; Chernobyl disaster viewed in, 19; deaths and illnesses related to Chernobyl disaster and, 19, 24–25; nuclear power stations in, 27; radioactive fallout in, 22–23; stability of, 27

Belov, 62

Beloyarsk nuclear power station, 131–32

Bikkenin, 173, 177

Bilibino, 13

Black market, 5

Blok, 88

Bogomolov, Volodya, 111

Boldyrev, V. A., 37, 101

Bolshakov, A. M., 113

Bolshoi Dom, 102

Bondarev, Yuri, 44

Bonner, Yelena, 165, 167–68

Borisova, Inna, 41–42, 43–44, 56, 156, 160

Brezhnev Constitution of 1977, 3

Brezhnev, Leonid Ilich, 14, 40, 74; nuclear power program under, 7–14; "period of stagnation" and, 3–7, 55–56, 71, 72–73, 110

Bryukhanov, Viktor, 20

Budker, 182

Bulgaria, 12

Bulgarian People's Army, 103

Bykov, Vasily, 43, 87, 95–96

Cancer: Chernobyl disaster and, 24, 25; radioactive fallout and, 41

Capitalism, in post-Soviet regime, 27, 28–29

Censors: secrecy vow of, 197; status of, 59; urged to write about censorship, 196, 197; writers and, 89–90

Censorship, 195–96; on articles on nuclear reactor accidents after Chernobyl disaster, 109–13, 131, 132, 145–46, 159; on articles on nuclear reactor accidents before Chernobyl disaster, 38–39, 41–42, 43–48, 56–59, 76, 77–82, 84–90, 91–104, 107–8, 180; banned words and, 90; beginning of the end of, 163, 177; Chernobyl nuclear power station disaster and, 109; on death in writing, 87; end of, 33, 38, 195; Gorbachev and, 167, 169–70, 197; KGB and, 43–45, 75, 79, 97–98, 103, 123–28; Kholchlova on, 33–36; ministries and, 37, 43; *perestroika* and, 157–58; process of, 59–60, 90; public onslaught against, 173–74, 175; radiation as consequence of, 38; secrecy regarding, 33, 35; Solntseva on, 36–37; of *The Truth About Chernobyl*, 136–52, 156–59, 161, 162–63,

164, 167, 171–72, 173–74, 178–80, 181–95. *See also* Glavlit; Gosatom; Nuclear censorship stamp; Secrecy

Center for Radiation Medicine, 18, 25

Chebrikov, Viktor, 189

Chelyabinsk nuclear power station, accident at, 2–3, 10, 41, 52, 58; secrecy on, 40, 83–84

Chernenko, 14, 85, 94–95, 97

Chernobyl commission, 136–37

Chernobyl Notebook (Medvedev), 160

Chernobyl nuclear power station, 11, 72; closure of, 26; defects in, 81; fuel melt in, 90; International Atomic Energy Agency and, 10; Pripyat at, 13, 20–21; as RMBK, 7

Chernobyl nuclear power station, disaster at, 1, 2, 3, 11, 29, 35, 83, 113; cause of, 20, 128; censorship and, 109, 141; Communist Party of Soviet Union and, 21–22; countdown to, 92, 93, 95, 99, 100; deaths related to, 16, 24, 25, 38; decontamination and reconstruction process after, 23–24; "The Expert Opinion" published after, 103; first days of, 14–20; as genocide, 130, 187; government response to, 16–18; illnesses related to, 18–19, 24–25, 128; Jewish conspiracy version of, 114–16, 178–80, 181–95; KGB and, 158; Kiev and, 22; lies and, 132, 164, 175; *Literaturna Ukraina* article and, 14–15; media reaction to, 15–16, 21; Medvedev's book on, 104; Medvedev's stories foreshadowing, 52–53; radioactive fallout from, 22–23, 25, 108, 112, 151; real story of, 20–25; reservists cleaning up, 23–24; sabotage and, 114–16, 178–80, 181–95; secrecy on, 15–20, 21, 22–23, 24, 116–36, 158, 177; Shcherbak's book on, 144; Shcherbina's diary on, 142; survivors of, 117; truth revealed on, 174–75. *See also Truth About Chernobyl, The*

Children of Chernobyl, 25

"China syndrome," 23

Chornovil, Vyacheslav, 7

Chusovoy, 69

Chyhyryn nuclear power station, 13

CIS. *See* Commonwealth of Independent States

Clinic No. 6, 39, 117, 128

Closed cities, 9

Coal industry: accidents in, 9; Brezhnev and, 4–5; in 1970s, 8–9

Commander, The (Karpov), 95

Commonwealth of Independent States (CIS), 26

Communism, in Soviet Union, 5

Communist Party of the Soviet Union (CPSU), 3; Brezhnev and, 5; elite of, 5

Communist Party of the Soviet Union (CPSU), Central Committee of: censorship and, 45, 78, 81, 100, 103, 157, 158, 163, 171, 175; cultural department of, 59; Jewish conspiracy version of Chernobyl disaster and, 190; nuclear power branch, 54; nuclear power industry and, 91; reaction to Chernobyl disaster, 21; on Soviet Writers' Union, 36

"Core, The" (Medvedev), in *Druzhba*, 86

Council of Ministers, 75, 91, 118, 128, 130, 138, 167, 171, 181

CPSU. *See* Communist Party of the Soviet Union

"Critical Path, The" (Medvedev), 81

Czechoslovakia, 6, 12

Davletbaev, Razim, 121, 130, 136

Death, censorship relating to writing about, 87

Dederichs, Mario, 23

Dedkov, Igor, 172–73, 177

Democracy, in post-Soviet regime, 27

"Departure from Nucleate Boiling Ratio, The" (Medvedev), 82, 84

Dimchevsky, Nikolai, 38, 40–41, 42, 43, 47–48, 109

Dissidents, 72; Brezhnev and, 6–7; KGB and, 75

Dnieper River, 13, 83

Dolotova, Natalia, 162, 165, 166, 168, 169

Don, The Reactor Unit in, 83, 159, 173, 180

Donbasenergostroy, 117

Donbass coal field, 9

Donetsk, 124

Dontekhenergo, 117, 123, 124

Druzhba, 86–87, 91, 114, 159; "The Core" in, 86; "The Expert Opinion" in, 100, 103, 107–8, 129, 134

Druzhba Narodov, 82, 84–85, 95

Dyatlov, 117

Dzurov, 103

Economy: Brezhnev and, 4–5, 6; Gorbachev and, 14; lack of nutritious food and, 25; Marshall Plan and, 9; "period of stagnation" and, 73; in post-Soviet regime, 27

Editors: censors and, 33; status of, 59

Electricity, from nuclear power stations, 12

Energodar, 13

Espionage, industrial, 132–34
"Expert Opinion, The"
    (Medvedev), 40, 94, 109, 129;
    brief account of, 52; censors'
    stamp for, 100; in *Druzhba*,
    100, 103, 107–8, 129, 134;
    reaction to, 134–35, 138

Finko, Oleg, 97–98
Firsov, Vladimir, 103, 159
Fominichna, Nina, 88, 89
Fomin, Mykola, 20
Food, nuclear power industry
    and, 54–55, 91
Frolov, Leonid, 143, 152

Gas industry, in 1970s, 8
Genocide: Chernobyl disaster as,
    130, 187; Stalin and, 130, 187
Genrikh, Oleg, 117, 121
Georgian Communists, 6
Ghandi, Rajiv, 163
Glasnost, 2, 18; censorship and,
    131; preserving appearance of,
    141; press and, 27
Glavatomenergo (Directorate for
    Nuclear Energy), 39
Glavlit (Directorate for Litera-
    ture), 33, 35, 36, 37, 47–48,
    58, 86, 88, 90, 97, 99, 101,
    108, 113, 137, 138, 145, 151,
    157, 158, 162, 163, 167, 172,
    175–76; beginning of end for,
    163; end of, 177–78; *The*

Truth About Chernobyl and,
    177
Glavstroy, 120, 123, 176
God, 101, 129; help of to victims
    of Chernobyl disaster, 130;
    Medvedev's dream about,
    147–48; Word as, 109, 129,
    160
Gomel, radioactive fallout in, 2,
    22–23
Gorbachenko, 117
Gorbachev, Mikhail S., 4, 7, 10,
    100, 137, 161, 163, 165, 167,
    168, 173; censorship and,
    197; on Chernobyl disaster,
    16; continuity of regime of,
    14; nuclear power program
    under, 14; Sakharov letter to,
    169–71
Gorky, 72
Gosatom, 38, 86, 88, 90, 92,
    93–94, 99, 131, 137, 139,
    140, 141, 144, 146, 151, 156,
    157, 158, 161, 163, 164, 167,
    171, 174; nuclear censorship
    stamp of, 43, 137–38, 140,
    143; secrecy on Chernobyl
    and, 177
Gosatomenergonadzor (State
    Committee on Operational
    Safety in the Nuclear Power
    Industry), 117, 140
Goskomgidromet (State Commit-
    tee on Hydrology and Meteo-
    rology), 132, 157, 164, 174
Gosplan, 75, 91

Graphite-moderated reactor. *See* RBMK

"Green Movement and Nuclear Power, The" (Medvedev), 61*n*

Grigorenko, Petro, 7

Grishin, Viktor, 14, 94

Gromyko, 150

Gubarev, Vladimir, 120, 192

Gulag Archipelago (Solzhenitsyn), 156

Heart attacks, among reservists cleaning up Chernobyl disaster, 24

Heat, reactors supplying, 13–14

"Hot Chamber, The" (Medvedev), 39, 40, 129; brief account of, 52

IAEA. *See* International Atomic Energy Agency

Ignalina nuclear power station, 11

Ignatenko, Yevgeny, 99–100, 101, 108, 122, 128–29, 137, 141, 150

Ilina, Valentina, 51

In August 1944 (Bogomolov), 111

"Incompetence" (Medvedev), 174, 175

Institute of Nuclear Physics, 186

International Atomic Energy Agency (IAEA), 10; nuclear power stations inspected by, 19; Soviet accounting of Chernobyl disaster to, 17

Isayev, Aleksandr, 86, 91, 107, 114, 115

Isayev, Lesha, 100

Ivanovna, Nina, 181, 193, 194

Ivanov, Vladimir, 123–28

Izrael, Yu. A., 174–75

Izvestia, 21, 37

Jewish conspiracy, Chernobyl disaster explained as, 114–16, 178–80, 181–95

Kalinin nuclear power station, 72

Kamenev, Yupiter, 84, 85, 99, 101, 108, 122, 150

Kamenyuga, Sergei, 89

Karlin, 57–59, 61

Karpov, Vladimir, 94, 95, 96, 109, 110–11

KGB, 56; Barkov and, 73–76; Brezhnev and, 3, 5; censorship and, 43–45, 75, 79, 97–98, 103, 123–28; Chernobyl accident secrecy and, 119–20, 158; Chernobyl disaster as sabotage and, 115, 116, 178–80, 181–95; dissidents and, 6–7; killing by, 87, 88; literary section of, 44–45, 87, 88; "period of stagnation" and, 72–73; at Politklub meeting, 134; response of to nuclear accidents, 90–91; *The Truth About Chernobyl* and, 117, 123–28

Khalyavchenko, Aleksei, 148–52

Khar'kiv nuclear power and heating station, 14

Khlopkov, Viktor, 74

Khmelnytsky nuclear power station, 12, 13

Khokhlova, Lilia, 33–36, 38, 90, 113, 151, 176, 178, 195

Khrushchev, Nikita, 3, 4, 5, 45, 47

Kiev, 15, 16, 18, 20, 21, 22, 83, 103, 184, 186

Kiev Communist party, 21

Kiev nuclear power and heating station, 14

Kiev Reservoir, 13

King Fish (Astafyev), 65

Klimovsky, Vladimir, 117, 119–20, 122

Kommunist, 47; *The Truth About Chernobyl* in, 42, 172–73, 174–75

Komsomol, 75, 79, 135, 189; Medvedev addressing, 134–35

Komsomolskaya Pravda, 36

Kontinent, 168

Kostroma nuclear power station, 11

Kostrov, Vladimir, 174

Kovalevska, Lyubov, 14–15

Krasnoyarsk, 66, 68, 71; accident at, 40

Krasnoyarsk-45, 58

Kravchuk, Leonid, 27

Kruglov, 179, 181

Kuleshova, Lyubov, 143–44

Kulov, Zhenya, 140–41

Kurchatov Institute of Atomic Energy, 11, 167

Kurgan, 65

Kurguz, 117

Kurochkin, Galina, 65

Kurochkin, Viktor, 65

Kursk nuclear power station, 11, 72

Larionov, Arseny, 81

Last Bow, The (Astafyev), 67

Laverov, 191

Legasov, Valerii, 17, 19–20

Lelechenko, 117

Leningrad nuclear power station, 7, 11, 113; accident at, 2, 40, 41, 52, 58

Lenin, Vladimir Ilyich, 3, 187

Leontiev, 134–35

Leukemia: Chernobyl disaster and, 24, 25; radioactive fallout and, 41

Library of the Molodaya Gvardia, 85–90

Ligachev, Egor, 17, 22, 107

Lipatov, Vilka, 69

Literaturna Ukraina, 14–15

Literaturnaya Gazeta, 44, 96

Literaturnaya Uchoba, 67, 91, 139, 185; "The Operators" in, 71, 76–81

Lithuania, 19
"Live and Remember"
    (Rasputin), 43
"Living Soul, A" (Medvedev), in
    Ural, 148
Lvov, Mikhail, 49, 56–57, 94,
    110

Maksamov, Lev, 178–80, 181–95
Maksimov, Vladimir, 168
Marin, Vladimir, 54, 132, 164,
    171, 172, 192
Markov, Georgi, 91, 110
Marshall Plan, 9
Mayorets, 152, 164, 180
Medvedev, Grigori, 1–3, 7, 10, 17,
    18, 19, 20, 25, 29; KGB in-
    quiring about, 123–28; on nu-
    clear power, 60–61, 66; prose-
    cution of, 36, 145, 148–52;
    writing process of, 67–68
Medvedev, Roy, 7
Medvedev, Vadim, 161, 167
Medvedev, Zhores, 10
Melekess, 101, 140
Metlenko, Gennady, 117–18,
    123, 124, 126
Mikhailov, Aleksandr, 67, 71, 91
Mikhalski, Vatslav, 113, 144–45
Military, display of under Brezh-
    nev, 5
Minatomenergo (Ministry of
    Atomic Energy), 19, 132, 138,
    141; nuclear censorship stamp
    of, 142, 159

Minenergo (Ministry of Energy),
    73, 117, 118, 126, 129, 176;
    "The Expert Opinion" and,
    134; nuclear censorship stamp
    of, 137
Ministries: Brezhnev and, 6; cen-
    sorship and, 35, 37, 43; octo-
    pus compared with, 35; Russia
    maintaining, 26
Ministry of Defense: censorship
    and, 103; on reservists at
    Chernobyl, 23
Ministry of Power, 8
Minsk nuclear power and heating
    station, 13–14
Minsredmash (Ministry of Medi-
    um Machine Building), 8, 58,
    86, 93, 101, 132, 133, 136,
    137, 139, 151, 179, 181, 185;
    "The Expert Opinion" and,
    134
Minvodkhoz (Ministry of Land
    Reclamation and Water Re-
    sources), 36, 143, 157
Minzdrav (Ministry of Public
    Health under Ilin), 164
Molodaya Gvardia, 85, 86, 89,
    93, 168
Molotov, 9
Moment of Life, A (Medvedev),
    114, 165; Sovietski Pisatel
    publishing, 148
Moscow, 16, 18, 21, 39, 69, 117,
    128, 148
Moscow Communist Party,
    21

Moscow Writers' Organization, 91

Moskovskie novosti (*Moscow News*), 28

Mykolaiv nuclear power station, 12

Nagibin, Yuri, 77

Narovchatov, Sergei, 48–51, 53, 56, 82, 87, 94, 180

Nash Sovremennik, 57–59, 168

Naumenko, Lev, 42, 44, 112

Neporozhny, 91

Netishyn, 13

Neva, 112, 131, 137, 139, 140, 156, 158, 162; "The Reactor Unit" withdrawn from, 145–46

Nikolayevich, Aleksandr, 184–85

Nikolsky, Boris, 112, 131, 137, 138, 139, 140, 141, 144, 145–46, 151, 158

No Breathing Room (Medvedev), 1–3, 8, 29

Nomenklatura, 5

Novosibirsk, 181, 182, 186, 189, 190, 192, 193; *Sem Dnyei* of, 114–16, 183–84, 185–86

Novosti, 185

Novovoronezh nuclear power station, 10, 72

Novy Mir, 41–42, 43, 48–51, 57, 59, 82, 94, 95, 97, 109, 155, 178, 179, 180, 181, 194; "The Reactor Unit" and, 110;

Sakharov and, 168; *The Truth About Chernobyl* and, 42, 143, 155–58, 160–65, 169, 171–73, 174–77

Nuclear censorship stamp, 85–86, 101, 137–38, 139–40; for collection of stories, 108–9; for "The Core," 86; for "The Expert Opinion," 100; of Glavlit, 177; of Gosatom, 43, 137–38, 140, 143; of Minatomenergo, 142, 159; of Minenergo, 137; for "The Operators," 80–81, 98–99; for "The Reactor Unit," 131, 138–41, 145–46, 148–51, 159; renewal of, 112; of Soyuzatomenergo, 86, 88–89, 99, 137, 150; for *The Truth About Chernobyl*, 143; *Ural* and, 101

Nuclear power, danger of. *See* Radioactive fallout

Nuclear power and heating stations. *See* ATETs

Nuclear power industry: control of, 8; grassroots movement against, 19; growth of, 8–14, 54, 71–72, 91; history of, 7–14; KGB and sabotage of, 74; Medvedev's position on, 60–61, 66; in post-Soviet regime, 26–27; rivers linked to, 13; secrecy regarding, 9–10; turnover in the labor force at construction sites in,

Nuclear power industry *(contined)*
55; in Ukraine, 12–14
Nuclear power stations, 73–74.
See also Accidents, in nuclear
power stations; Chernobyl nuclear power station; *under specific power stations*
"Nuclear Tan, A" (Medvedev),
82, 87, 91, 95–97, 101, 109,
129; reaction to, 131–32; in
Ural, 96, 131–32

Odessa nuclear power and heating station, 13
Odoyevsky, 92, 93, 139, 140
Ogonyok, 168
Oil: Brezhnev and, 4; 1970s and,
8; nuclear power industry and,
54–55, 91
*One Day in the Life of Ivan Deniso-
vich* (Solzhenitsyn), 45, 50
Onezh Lake, 63
Operators, The (Medvedev), 139;
publication of, 97–99, 136
"Operators, The" (Medvedev),
39, 40, 43, 53, 56, 61, 91, 93;
Astafyev and, 66–67, 71; brief
account of, 52, 53; in *Literaturnaya Uchoba,* 71, 76–81; nuclear censorship stamp for, 80
Ovsyanka, 68

Palamarchuk, 121
Peaceful atom, 54; danger of,

83–84, 92–93; need to write
about, 55, 69; "The Operators" and, 76; "The Reactor
Unit" and, 82
Perestroika, 2, 18, 60, 95, 190;
censorship and, 131, 157–58,
163; Chernobyl disaster and,
116, 190; "The Expert Opinion" and, 103, 134; industrial
espionage and, 133–34;
"period of stagnation" and,
73
"Period of stagnation," 3–7,
55–56, 71, 72–73, 110
Petrosian, 167
Petrov, 117
Petrozavodsk, 63
Pikul, Valentin, 58
Poland, 12
Politburo, 14, 17, 21, 45, 161,
163, 167, 171, 172, 191, 192
Politklub, Medvedev addressing,
134–35
Post-Soviet regime: capitalism in,
27, 28–29; democracy in, 27;
economy in, 27; nuclear accidents in, 2, 26; nuclear power
industry in, 26, 27
Prague Spring, 6
Pravda, 114, 186, 188
Press: on Chernobyl disaster, 16,
21; *Izvestia,* 21, 37; *Liturnia
Ukraina,* 14–15, 27;
Moskovskie novosti, 28; in
post-Soviet regime, 27–28;
Pravda, 114, 186, 188; *Sem*

Press *(continued)*
  Dnyei, 114–16, 183–84,
    185–86; TASS, 16
Pripyat, 13, 14, 20–21
Pripyat River, 120
Protection, as banned word, 90
Prushinsky, Boris, 108, 121, 122,
    131, 150
Pushkin, 137

Radiation sickness, 39–40, 42,
    69; from Chernobyl disaster,
    128
Radioactive fallout, 52–53; cen-
    sorship related to, 38; from
    Chernobyl disaster, 21–25, 38,
    83, 90, 108, 112, 151; deaths
    from, 41; in Krasnoyarsk, 68,
    71; radiation sickness from,
    39–40, 42, 69, 128
Rakov, Aleksei, 86, 87–88, 92,
    94, 98, 100, 115, 134–35,
    138, 158–60
Rasputin, Valentin, 43–44
RBMK (graphite-moderated re-
    actor), 7–8, 11–12; Chernobyl
    nuclear power station as, 7,
    11; diagram of in *Sem Dnyei*,
    114–16, 183–84, 185–86; mil-
    itary control over, 10; safety
    of, 11–12
"Reactor Unit, The" (Medvedev),
    10, 40, 41, 43, 44, 48, 49, 53,
    58, 61, 82, 83, 84, 95, 97,
    109–12, 129, 131, 138–41,

    162, 183; Astafyev and, 66–67;
    brief account of, 51–52; cen-
    sorship of, 145–46; in *Don*, 83,
    159, 173, 180; investigation of,
    162; Medvedev's prosecution
    and, 145, 148–52; in *Novy
    Mir*, 110; nuclear censorship
    stamp for, 138–41, 159; reac-
    tion to, 159–60
Reshetnikov, Yevgeny, 117, 120,
    141–42, 159
"Right to a Name, The"
    (Bykov), 96
River systems, nuclear power in-
    dustry and, 13
Romanenko, Anatolii, 25
Romania, 12
Rostov-na-Donu, 83, 159, 173
Rostov nuclear power station,
    173
Rovenskaya nuclear power station,
    71–72, 123; accident in, 82
Rubtsov, Nikolai, 62–63, 70
Russia, 27; nuclear power indus-
    try in, 27; stability of, 27;
    Ukraine and, 26. *See also* Post-
    Soviet regime
Ryzhkov, Nikolai, 17, 22, 180

Sabotage: Chernobyl disaster ex-
    plained as, 114–16, 178–80,
    181–95; KGB and, 74, 90, 91
Safety: as banned word, 90;
    Chernobyl nuclear power
    station and, 81; secrecy on

Safety *(continued)*
nuclear power stations regarding, 11–12, 13, 71–72
Safonov, Ernst, 67
Sakharov, Andrei, 7, 45, 72, 79, 162, 164–71, 175, 189
Samchenko, Vitaly, 132–33
Sarcophagus, reactor unit No. 4 enclosed with, 120
Sarcophagus (Gubarev), 120
Satsky, Aleksandr, 67
Savinich, Sergei, 176, 177
Savushkina, Tatiana, 117, 118, 122–28
Secrecy: on censorship, 33, 35; on Chelyabinsk explosion, 10; on Chernobyl disaster, 15–20, 21, 22–23, 24, 116–36, 158; on domestic industry, 9; Medvedev violating principle of, 102; on Ministry of Medium Machine Building, 8; on nuclear accidents in post-Soviet regime, 2; on nuclear power industry, 9–10; on nuclear power plant accidents, 40, 44, 47, 78, 80, 82, 83–84; on safety of nuclear power stations, 11–12, 13, 71–72. *See also* Censorship
Sem Dnyei, RBMK reactor diagram in, 114–16, 183–84, 185–86
Sergeyeva, Inna, 82–83
Severnaya thermal power station, 35, 36

Shaginyan, Marietta, 44, 45, 46–47
Shasharin, Gennady, 121, 122
Shcharansky, Anatolii, 7
Shcherbak, Yuri, 6, 20, 144, 192
Shcherbina, Boris, 128, 130, 131, 132, 136, 142, 144, 146, 152, 158, 164, 167, 171, 177, 180
Shcherbytsky, Vologymyr, 14
Shishkin, Vladimir, 121
Shugalo, 126, 127
Shushkevich, Stanislav, 27
Skachkov, Igor, 146
Slezko, 100
Slyunkov, Nikolai, 22, 190, 191, 192, 194
Smagin, Viktor, 121, 130, 136
Smirnov, 140
Smolensk nuclear power station, 11, 72, 81, 108
Socialism, in Soviet Union, 5
Sokolov, Dmitri Dmitriyevich, 93–94, 102
Solntseva, Natalia, 36–37
Solodin, Vladimir, 145, 162–63, 175, 176, 177
Solzhenitsyn, Aleksandr, 7, 50, 78, 156
Sosnovy Bor, accident at, 40, 52. *See also* Leningrad nuclear power station
Soviet Academy of Sciences, Novosibirsk branch of, 182, 186, 192
Soviet Congress of People's Deputies, 168

Sovietskaya Kultura, 120
Sovietskaya Rossiya, 41, 81, 109
Sovietskaya Rossiya publishing
  house, 38, 40, 41
Sovietski Pisatel, 37, 113–14,
  143, 148; *The Truth About
  Chernobyl* and, 144–45,
  146–47
Soviet Writers' Union, 36, 48,
  79, 91–92, 110
Sovremennik publishing house,
  85, 97, 98, 108, 136, 139,
  152; *The Truth About Cher-
  nobyl* and, 143–44
Soyuzatomenergo (All-Union
  Department for Nuclear Ener-
  gy), 73, 84, 99, 108, 120, 121,
  122–23, 124, 128, 131; nu-
  clear censorship stamp of, 86,
  88–89, 99, 137, 150
Soyuzatomenergostroy (All-
  Union Department for the
  Construction of Nuclear
  Power Stations), 73, 121
Soyuzelektromontazh (All-Union
  Department of Electrical As-
  sembly), 121
Soyuzglavzagranatomenergo (All-
  Union Directorate for Foreign
  Nuclear Power), 101
Soyuztsentratomenergostroy (All-
  Union Central Department for
  the Construction of Nuclear
  Power Stations), 73
Stakhanov, 10
Stakhanovism, 14

Stalin, Joseph, 9, 14, 28, 88, 137;
  censorship and, 34; genocide
  by, 130, 187
Stepanov, Yuri, 80
Straraya Ruza, 57
Strelnikov, Misha, 69
Strelyany, Anatoly, 143
Suslov, Mikhail, 84–85, 95
Sverdlovsk bacterial weapons
  plant, accident at, 45–46
"Syndrome, The" (Medvedev),
  40, 82, 95–96

TASS, on Chernobyl disaster, 16
Teplukhin, Benyamin, 76–79, 80,
  185
Ter-Akopian, 83
Tevekelyan, Diana, 44, 50, 51,
  53–54, 56, 97
Three Mile Island, 17, 82
Tikhomirov, 185, 189, 191, 192
Timofeyeva, Margarita, 155, 156,
  160, 162
Tomsk nuclear power station, ac-
  cident at, 40, 100
Toptunov, 187
Tormozin, 121
Towns, growth of along with nu-
  clear power stations, 12–13
"Trail of the Inversion, The"
  (Medvedev), 42
Trifonov, Yuri, 85
Truth About Chernobyl, The
  (Medvedev), 1, 39, 107, 113,
  136; censorship of, 136–52,

Truth About Chernobyl (continued)
156–59, 161, 162–63, 164,
167, 171–72, 173–74, 178–80,
181–95; defense of, 176–77;
eye witnesses of the events for,
116–18, 121–22, 130–31;
KGB and, 123–28; in *Kom-
munist*, 42, 172–73, 174–75;
Novy Mir and, 42, 143,
155–58, 160–65, 169, 171–73,
174–77; Sakharov and, 45,
162, 164–71, 175; secrecy of
information for, 116–36; and
Sovietski Pisatel, 144–45,
146–47; and Sovremenik,
143–44; title of, 160
Tsvirko, Mikhail, 121, 122
Tvardovsky, Aleksandr, 7, 50

Ukraine, the, 19; Chernobyl disaster viewed in, 19; Donbass
coal field in, 9; illnesses and
death after Chernobyl disaster
in, 19, 24, 25; nuclear power
industry in, 12–14, 15, 26, 27;
Russia and, 26; Shcherbytsky
and, 14; stability of, 27
Ukrainian Communist party, 21
Ukrainian Ministry of Health,
24
Ukrainian Popular Movement
(*Rukh*), 27
Ukrainian Soviet Socialist Republic, 114, 184

Ukrainian Writers' Union, 14
Ural, 85, 101, 109, 148; "A Nu-
clear Tan" in, 96, 131
Ural Mountains, 10
USSR Writers' Union, 174
Ustinov, D., 94
Uzunov, 185

Velikhov, 190
Verkhovykh, 136–37, 141, 146,
164
Victory Day celebrations, 61,
63
Vidrashka, Feodosy, 59, 143,
175–76, 181, 193, 194
Vlasin, 188
VNIIAES (All-Union Scientific
Research Institute for Nuclear
Power Stations), 118
Vodolazhko, 117
Voice radio stations, 89
Volga, 5, 80–81, 135, 179, 181
Volgodonsk, 12
Vologda, 61, 62, 68
Vologda River, 62
Voronin, Leonid, 118, 120
Voronov, Vasily, 159, 173
VVER, 12, 13; at Novovoronezh,
10

Word: "The Expert Opinion"
and, 138; as God, 109, 129,
160

World Festival of Youth and Students, The, 89
Writers: censors and, 33, 89–90; status of, 59–60

Yakovlev, Aleksandr, 157, 161, 167, 172
Yaroslavtsev, Boris, 101–2
Yefimov, Andrei, 98–99, 108
Yeltsin, Boris, 26, 27
Yeremeyev, N. I., 92–93, 139
Yermolova, 66
Yid-Mason version, of Chernobyl disaster, 114–16, 178–80, 181–95

Yuzhno-Ukrainskaya nuclear power station, 72, 117, 124; accident at, 90

Zalivaka, Valery, 144
Zalygin, Sergei, 43, 143, 155, 156, 157, 158, 161, 162, 163, 164, 165, 167, 168, 171–72, 174, 175, 177, 178, 181, 183, 185, 193
Zapadnaya Litsa, 50
Zaporozhiye nuclear power station, 13, 74, 90
Zayets, 122
Zhukov, Anatoly, 97, 103, 110